Proceedings of the Institute of General Physics
Academy of the Sciences of the USSR
Series Editor: A.M. Prokhorov
Volume 3

THE MAGNETIC AND ELECTRON STRUCTURES OF TRANSITION METALS AND ALLOYS

Edited by
V.G. Veselago and L.I. Vinokurova

Translated by Paul Makenin
and Kevin S. Hendzel

NOVA SCIENCE PUBLISHERS
COMMACK

NOVA SCIENCE PUBLISHERS
283 Commack Road
Suite 300
Commack, New York 11725

This book is being published under exclusive English Language rights granted to Nova Science Publishers, Inc. by the All-Union Copyright Agency of the USSR (VAAP).

Copyright 1988 Nova Science Publishers, Inc.

Printed in the United States of America

Proceedings of the Institute of General Physics
Academy of the Sciences of the USSR
Series Editor: A.M. Prokhorov

Proceedings of the Institute of General Physics
Academy of the Sciences of the USSR
Series Editor: A.M. Prokhorov

Volume 1 **Oceanic Remote Sensing,** Edited by F.V. Bunkin and K.I. Volyak.

Volume 2 **Laser Raman Spectroscopy in Crystals and Gases,** Edited by P.P. Pashinin.

Volume 3 **The Magnetic and Electron Structures of Transition Metals and Alloys,** Edited by V.G. Veselago and L.I. Vinokurova.

Volume 4 **Laser Methods of Defect Investigations in Semiconductors and Dielectrics,** Edited by A.A. Manenkov.

Volume 5 **Fiber Optics,** Edited by E.M. Dianov.

Volume 6 **The Nonlinear Optics and Acoustics of Fluids,** Edited by F.V. Bunkin.

Volume 7 **Formation and Control of Optical Wavefronts,** Edited by P.P. Pashinin.

Volume 8 **Problems of Lithography in Microelectronics,** Edited by T.M. Makhviladze.

Volume 9 **Selective Laser Spectroscopy of Activated Crystals and Glasses,** Edited by V.V. Osiko.

CONTENTS

MAGNETIC T–c PHASE DIAGRAMS FOR ORDERED IRON–PLATINUM AND IRON–RHODIUM ALLOYS WITH COMPOSITIONS IN THE TRANSITION REGION FROM ANTIFERROMAGNETISM TO FERROMAGNETISM

L. I. Vinokurova, A. V. Vlasov, V. Yu. Ivanov, M. Pardavi-Horváth, and E. Schwab

INSTABILITY OF MAGNETIC STATES IN ORDERED IRON–PLATINUM AND IRON–RHODIUM ALLOYS

V. G. Veselago, L. I. Vinokurova, A. V. Vlasov, N. I. Kulikov, B. K. Ponomarev, M. Pardavi-Horváth, and L. I. Sagoyan

KINETIC PROPERTIES OF IRON–RHODIUM ALLOYS IN THE TRANSITION REGION FROM ANTIFERROMAGNETISM TO FERROMAGNETISM

L. I. Vinokurova and V. Yu. Ivanov

MAGNETIC PROPERTIES OF THE ALLOYS $Pt_3Mn_xFe_{1-x}$ AND $(Pd_xPt_{1-x})_3Fe$ IN STRONG MAGNETIC FIELDS

V. Yu. Ivanov, Yu. N. Tsiovkin, N. I. Kourov, and N. V. Volkenshtein

FERMI SURFACES OF MAGNETIC TRANSITION METALS UNDER PRESSURE

A. G. Gapotchenko, E. S. Itskevich, and É. T. Kulatov

LOCALIZED MAGNETIC MOMENTS IN METALS AND ALLOYS WITH SPIN-DENSITY WAVE INSTABILITY

E. T. Kulatov, N. I. Kulikov, and V. V. Tugushev

vii

A STUDY OF $Zn_xCd_{1-x}Cr_2Se_4$ SPIN GLASSES
A. V. Myagkov and A. A. Minakov

MAGNETIC T–c PHASE DIAGRAMS FOR ORDERED IRON–PLATINUM AND IRON–RHODIUM ALLOYS WITH COMPOSITIONS IN THE TRANSITION REGION FROM ANTIFERROMAGNETISM TO FERROMAGNETISM

L. I. Vinokurova, A. V. Vlasov, V. Yu. Ivanov, M. Pardavi-Horváth, and E. Schwab

Abstract Detailed research was carried out on the magnetic properties of iron–platinum and iron–rhodium alloys with compositions in the transition region between antiferromagnetism and ferromagnetism. It was found that short-range magnetic order (including "cluster-glass-embedded-in-an-antiferromagnetic-matrix"-type ordering) is established in the alloys when the iron concentration exceeds that in the stoichiometric alloys Pt_3Fe and FeRh.

INTRODUCTION

Ordered iron–platinum alloys with compositions close to the stoichiometric composition Pt_3Fe and iron–rhodium alloys with compositions close to the equiatomic composition FeRh exhibit many behavioral features in common.

Previous magnetic, neutron diffraction, and Mössbauer studies [1, 2, 3, 4] have indicated that ordered iron–platinum alloys (Cu_3Au-type ordering) with stoichiometric composition are antiferromagnetic at low temperatures; as the iron content of the alloys increases, a composition-induced phase transition to a ferromagnetic state occurs at approximately 36 at%. However, the magnetic structure of the alloys for compositions in the transition region from antiferromagnetism to ferromagnetism and, even more so, the way in which these structures are formed has remained an open question.

At low temperatures, Fe–Rh alloys of nearly equiatomic composition (iron concentration $48 \lesssim c_{Fe} \lesssim 51$ at%) are oriented antiferromagnetically with a G-type structure described by the wave vector $\mathbf{k} = [1/2, 1/2, 1/2]$ [5,6]. A first-order phase transition into a ferromagnetic state is observed at a temperature of approximately 320 K; this ferromagnetic state holds until the Curie temperature $T_c \sim 670$ K [7]. The presence of this transition is what has given rise to the enormous interest

1

in research on Fe–Rh alloys. In spite of the large number of experimental and theoretical papers on the properties of Fe–Rh alloys, the magnetic structures of the alloys as a function of composition have not yet been determined. All that is known is that they become ferromagnetic at an iron concentration $c_{Fe} = 51.5$–52.0 at% over the entire temperature interval where the magnetically ordered state exists.

In this paper, we present the results of a detailed study of the properties of iron–platinum and iron–rhodium alloys with compositions in the transition region from antiferromagnetism to ferromagnetism. This study was aimed at measuring the magnetic phase transitions that occur in the alloys at various compositions and temperatures.

Some data on the magnetic properties of iron–platinum alloys have been published in [8].

1. PREPARATION OF THE SAMPLES

Pt–Fe alloy monocrystals were obtained at the Institute for Metal Physics (Ural Science Center, USSR Academy of Sciences). The alloys were prepared using 99.9% pure platinum and carbonyl iron smelted in a vacuum furnace.

The alloys were smelted in an induction furnace with an inert gas atmosphere (0.3–0.5 atm gauge pressure). After smelting twice, the ingots were smelted once again in a tungsten-electrode arc furnace in an inert atmosphere and poured into water-cooled copper molds to obtain 3–5 mm-diameter bars that were then used as stock for growing the monocrystals. The chemical compositions of the smelted alloys are given in Table 1.

The monocrystals were grown by the Bridgman method in cylindrical alundum crucibles with conical bottoms (opening angle 30°) in an argon atmosphere. The

Table 1. Chemical Composition of Pt–Fe Alloys

Alloy	Fe concentration, mass%	
I	8.66	(24.90)
II	9.46	(26.76)
III	11.47	(31.04)

NOTE: Fe concentration in at% in parentheses.

crystallization speed was 1 mm/min. The resulting monocrystals were 6–7 mm in diameter and 30–50 mm in length. The monocrystals were annealed for 50 hr in an argon atmosphere at 1400–1420 °C, and subsequently cooled slowly in the furnace (3–5 hr).

The cross-sectional homogeneity of the monocrystal chemical composition was checked in Fe K_α radiation using a Microscan IIA microanalyzer. The properties of the cross-sectional distribution of iron across the crystal indicate that it is extremely homogeneous. However, the iron concentration increased somewhat along the axis of crystal growth, from the region of the onset of crystallization to the region where crystallization ended, so that samples cut from different regions in the crystal differed in composition by anywhere from a few tenths of a percent to 1–2 at% Fe.

An SN-144 electric arc cutter was used to cut the samples, which were mounted in a special goniometer used to set the spatial orientation of the monocrystal before cutting. The monocrystals were cut along either the (001) or (110) planes into templets 0.7 mm thick. Both parallepipedal samples (with mean dimensions of 0.7 × 0.7 × 7 mm having a particular crystallographic axis oriented along longitudinally) and cubes of side length 2.7 mm were cut from the templets.

The crystallographic orientation of the samples was determined by X-ray crystallography (white Mo radiation) using an RKSO-2 camera.

To remove the effects of stress resulting from the electric arc cutting procedure, the samples then underwent stepwise annealing in sealed quartz ampules for 20 hr at 670°C, 20 hr at 600°C, and 10 hr at 500°C, and were then cooled slowly in the furnace to room temperature at a rate of 0.7°/min.

The presence of superlattice reflections in the Laué diffraction patterns indicated that all of the samples had ordered structure. The degree of large-scale ordering determined by X-ray crystallography from the intensity ratios of the (100) and (110) superlattice reflections and the (200) and (220) lattice reflections was equal to unity.

The iron–rhodium alloy samples were prepared at the Central Research Institute for Physics of the Hungarian Academy of Sciences by electron-beam smelting in an argon atmosphere. The samples were prepared using (99.99%) iron and (99.996%) rhodium. These samples were then heat-treated/annealed for 4 hr in a pure hydrogen atmosphere at $T = 1000°C$ and hardened in air. Research samples with various iron concentrations ranging from 48.0–52.0 at% were prepared. X-ray crystallography indicated that the samples were ordered, and had CsCl-type structure.

1. MAGNETIC PHASE DIAGRAM FOR THE Pt–Fe ALLOYS

1.1. DIFFERENTIAL MAGNETIC SUSCEPTIBILITY

The differential magnetic susceptibility χ_d was measured inductively in an alternating current with frequencies ranging from 120 Hz to 1.2 kHz and a variable magnetic field amplitude ranging from a few tenths of an Oersted to several Oersted. The difference signal from a compensated input coil was amplified by a U2-8 narrow-band amplifier and then fed, after synchronous or linear detection, into the Y axis of a PDS-021 recorder whose X-axis was fed the signal from a thermocouple. The measurements were mainly carried out on parallepipedal samples having a low demagnetization factor ($N \approx 0.1$. A weak, constant magnetic field was applied parallel to the sample axis in several cases.

In studying the weakly magnetized materials, the variable signal from the frequency output of the U2-8 amplifier was fed into a V3-40 voltmeter. The rectified and amplified voltage, which was of order a few hundred millivolts, was compensated by a constant voltage obtained from a voltage divider or a R-307 high-resistance potentiometer, and only a portion of the voltage, of order a few tens of millivolts or even a few millivolts was fed into the Y-axis of the recorder. This method allowed us to reliably record a useful signal that was less than 5% of the total background, or to record changes in susceptibility of order 10^{-4}.

Figure 1 shows the functions $\chi_d(T)$ for alloys with compositions in the transition region.

The figure shows that all of the alloys studied have anomalies in χ_d. Similar anomalies in susceptibility are also observed in relatively weak constant magnetic fields (Curve 4 in Fig. 1a). In what follows, we shall denote the temperature of the low-temperature anomaly by T_1 and that of the high-temperature anomaly by T_2.

In alloys with $c_{Fe} < 30.5$ at%, the anomaly at T_1 takes the form of a sharp jump with approximately 1 K hysteresis; this is indicative of the presence of a first-order phase transition. Application of an external magnetic field shifts this jump to lower temperatures, without a change in the size of the jump. However, the field has practically no effect on the position of the high-temperature peak at $T = T_2$, which rapidly decreases in size with increasing field strength, and is completely suppressed by $H_L = 200$ Oe. At higher field strengths, only a change in slope of the $\chi_d(T)$ curve is observed at $T = T_2$.

The function $\chi_d(T)$ for the alloy with $c_{Fe} = 30.5$ at% has only a single maximum at $T = 124$ K; this maximum decreases in size when an external magnetic field is applied.

Two closely spaced anomalies are once again observed in the alloys containing 30.65 at% Fe. The anomalies become more widely separated when an external field (which shifts the anomalies in opposite directions) is applied.

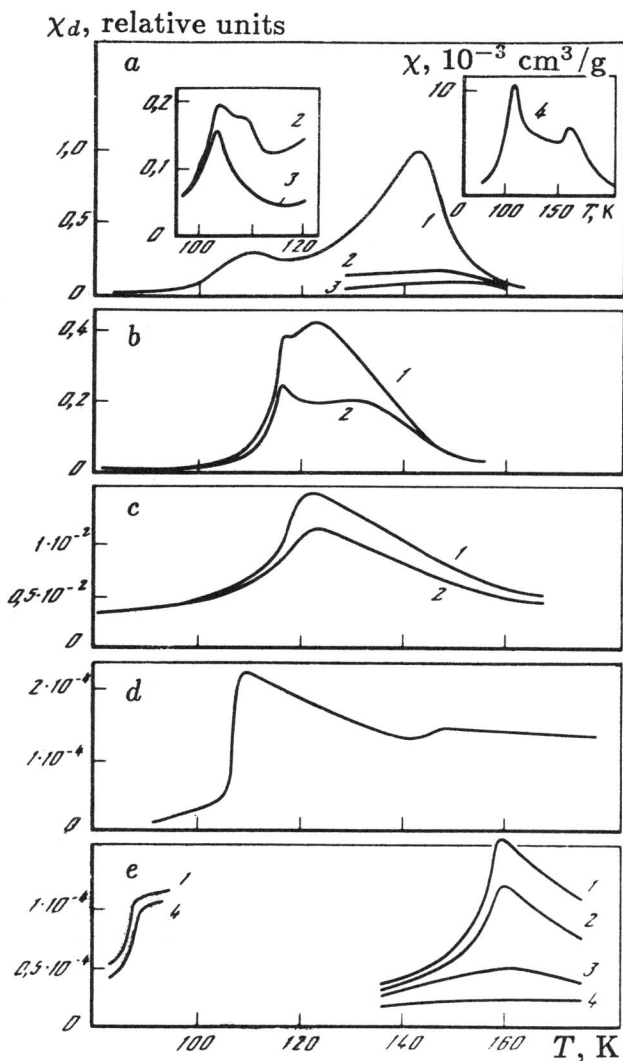

Figure 1 Differential magnetic susceptibility as a function of temperature for iron–platinum alloys (the amplitude of the variable magnetic field in the primary coil was 0.3 Oe). (a) 32.16 at% Fe ((1) $H_L = 0$, (2) 40, (3) 80 Oe, (4) measurement of χ in a constant field $H = 235$ Oe); (b) 30.65 at% Fe ((1) $H_L = 0$, (2) 18 Oe); (c) 30.5 at% Fe ((1) $H_L = 0$, (2) 20, (3) 80, (4) 200 Oe).

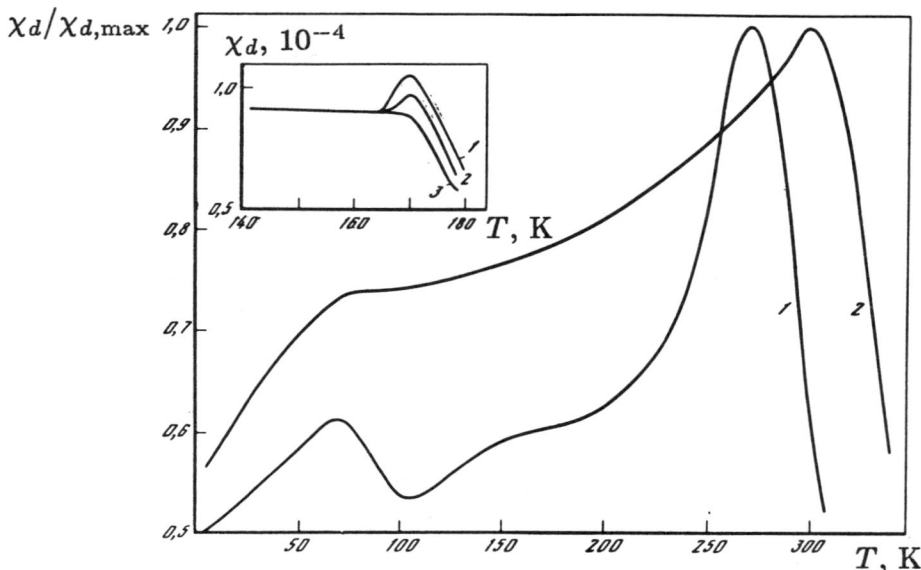

Figure 2   Differential magnetic susceptibility as a function of temperature for iron–platinum alloys with $c_{Fe} = 34$ (1) and 36 at% (2) $(H_L = 0)$. Inset: $c_{Fe} = 24.8$ at%: (1) $H_L = 0$; (2) 16; and (3) 32 Oe.

The samples with $c_{Fe} > 32$ at% have an additional anomaly between T_1 and T_2 at $T = T_3$ $(T_3 = 113$ K for $c_{Fe} = 32.16$ at%). This peak decreases in size and shifts to lower temperatures as the external field increases, and becomes indistinguishable against the background of the peak at $T = T_1$.

Figure 2 shows the differential magnetic susceptibilities for alloys with nearly stoichiometric composition $(c_{Fe} = 24.8$ at%) and the alloys with iron concentrations of 34 and 36 at%. The alloys with $c_{Fe} = 24.8$ at% are observed to have only one peak χ_d at $T = 170$ K, which is rapidly suppressed by an external magnetic field. The alloys with iron concentrations of 34 and 36 at% have a previously unobserved anomaly in χ_d at $T \approx 80$ K in addition to the prominent high-temperature anomalies.

Similar anomalies were also observed in the temperature dependences of the electrical resistance, galvanomagnetic effects, heat capacity, coefficient of thermal expansion, and magnetostriction (these properties will be discussed in other pa-

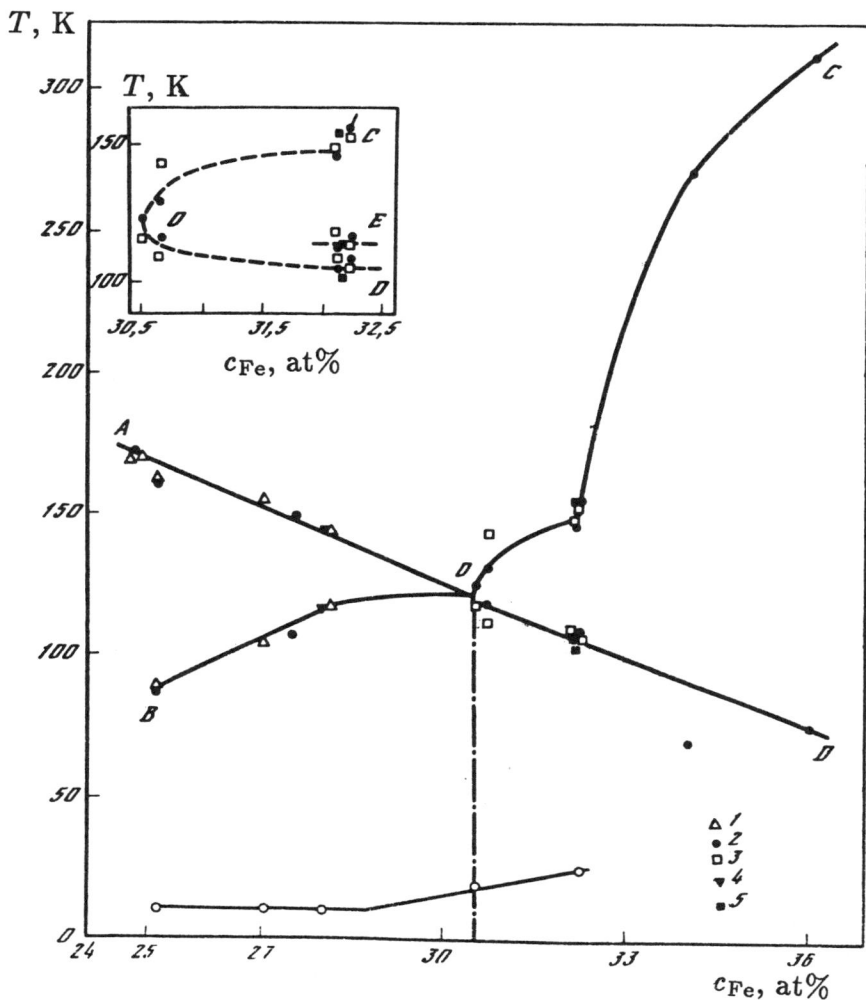

Figure 3   Magnetic phase diagram for ordered iron–platinum alloys with compositions in the transition region from antiferromagnetism to ferromagnetism. Transition temperatures determined: (1) from anomalies in the electrical resistance; (2) from maxima in the differential magnetic susceptibility; (3) from maxima in the magnetic reluctance; (4) from anomalies in the heat capacity; and (5) from neutron diffraction studies. Open circles: freezing point for cluster glass. Inset: enlarged phase diagram for alloys containing 30.5–32.2 at% Fe.

pers in this volume). The observed anomalies were used to construct a magnetic phase diagram for Fe–Pt alloys with compositions in the transition region between antiferromagnetism and ferromagnetism (Fig. 3).

1.2. NEUTRON DIFFRACTION STUDIES

Coherent elastic neutron scattering was used to determine the types of large-scale magnetic ordering in the vicinity of the magnetic phase transitions for three monocrystalline Pt–Fe alloys containing 32, 30.5, and 28 at% Fe over a temperature range from 4.2 to 290 K. The study was carried out with a diffractometer on the horizontal reactor channel at the Central Research Institute for Physics of the Hungarian Academy of Sciences. The neutrons, which were formed by reflection from the (002) plane of a zinc monocrystal monochromator, had a wavelength of 1.06 Å. The contamination from second-order neutrons was approximately 1.08%.

The scanning over angle (every 0.05°) was carried out automatically for each temperature. The angles and scattered neutron intensities were recorded on paper tape for later computer processing and printed out on a line printer. The integrated scattered neutron intensity was calculated for the extrapolated Gaussian curves for the experimental angular intensity distributions of the magnetic and nuclear reflections. The magnetic moments were calculated using various probable models for the magnetic structure of the alloys. The temperature was set and controlled to within 0.1 K.

Alloys with $c_{Fe} \geq 30.5$ at% Fig. 4 shows the intensities of some of the reflections for alloys with $c_{Fe} = 32$ at% as a function of temperature. In agreement with the magnetic measurements, the neutron diffraction studies indicated that the alloys had three distinct structures in the temperature region from 4.2 K–150 K.

In complete agreement with the low-temperature results of Bacon [1] and Men'-shikov [2], other than the nuclear reflections, only reflections of the type {1/2 0 0}, {1 1/2 0}, and {3/2 0 0} corresponding to antiferromagnetic structure with wave vector $\mathbf{k} = [1/2, 0, 0]$ were observed. This structure can be characterized by the Lifshits star $\{\mathbf{k}_{10}\}$ in the simple cubic lattice Γ_c formed by the iron atoms at the vertices of the cube [9].

The reflections from crystal faces of the same type, for example (1/2 0 0), (0 1/2 0), or (0 0 1/2), turned out to be of approximately the same strength to within 5% (the accuracy with which the crystal was aligned). This may occur for a crystal with multiple domains or when the structure is noncollinear and there is a multiple-beam path. The path along the $\{\mathbf{k}_{10}\}$ star can be determined from the intensity ratios of the various reflections, both in studies of powder samples and in studies of monocrystals. We concluded, in agreement with the earlier results obtained for powders [1, 2], that the Pt–Fe alloy has a single-beam path. The isostructural ternary alloys $Fe(Pd_{1-x}Au_x)_3$ [10] and $Fe(Pt_{0.46}Pd_{0.52}Fe_{0.02})_3$ [11] also have a single-beam path along the $\{k_{10}\}$ star.

Magnetic moments were calculated both for approximate models in which only the magnetic moments on the iron atoms at the vertices of the cubical cells were taken into account as well as in the more general case, where the contributions to the coherent neutron scattering from "additional" iron atoms on the faces and on ferromagnetic "sheets," as well as from platinum atoms on opposite faces, were taken into account. Such a calculation was carried out by Men'shikov, et al. [2], who found that $\mu_{Pt} \approx 0.2\mu_B$, with the same orientation as the iron atoms on a given "ferromagnetic sheet." The formulae in [2] assume that the presence of an "additional" iron atom on a given face uniquely determines

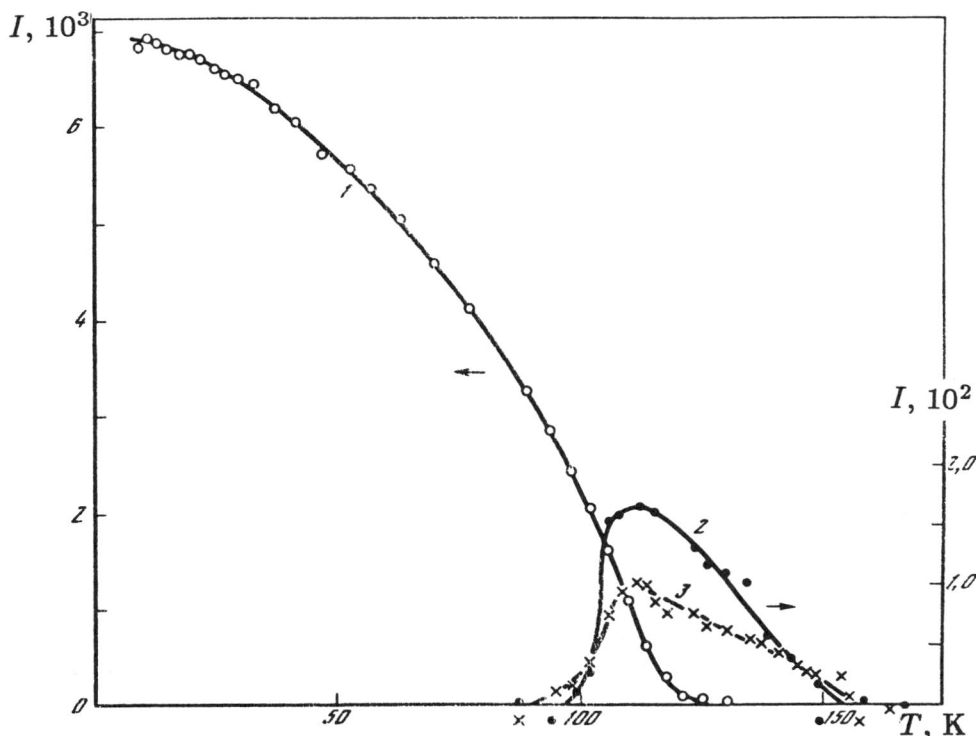

Figure 4 Intensity of elastic neutron scattering from various crystallographic planes in $Pt_{6.8}Fe_{3.2}$. (1) (1/2 0 0); (2) (1 0 0); (3) (1 1 0).

the formation of a ferromagnetic "sheet" on that face. However, one can also imagine another situation in which the "additional" iron atoms are distributed with equal probability between the crystal faces, i. e., so that only one third of them fall on "ferromagnetic sheets."

The calculations indicated that best agreement with the experimentally measured intensities is obtained when the moments of the "additional" iron and platinum atoms are taken into account; even though the magnitude and direction of the magnetic moment on the iron atoms turned out to be relatively insensitive to the choice of model (at 4.2 K, the magnetic moments of the iron atoms $(\mu_{Fe} \approx (3.3 \pm 0.3)\mu_B)$ are oriented perpendicular to the wave vector $\mathbf{k}$ and distributed over equivalent directions, such as (for example) [010] and [001] when $\mathbf{k}\|[100]$, with uniform probability), the orientations of the magnetic moments on the platinum atoms $(\mu_{Pt} \lesssim 0.3\mu_B)$ are found to be either parallel or antiparallel to μ_{Fe} in calculations for the two extreme cases for the probability distribution of the "additional" iron atoms, so that polarized neutron experiments are needed for final determination of the magnitude and direction of the magnetic moments on the platinum atoms.

No anomalies are observed in the coherent neutron scattering as a function of temperature at low temperatures $(T < T_1)$, where anomalies are observed in the magnetic and kinetic properties of the alloys (see Section 2 and subsequent papers in this volume).

As the temperature increases to $T \approx 100$ K, the antiferromagnetic reflections are accompanied by {100} and {110} reflections due to the ferromagnetic structure characterized by the star $\{\mathbf{k}\} = 0$. Such structures were only observed on the temperature interval from 100 to 112 K, i. e., in the region between the low-temperature anomalies in χ_d and $\Delta\rho/\rho_0$. Since no broadening of the reflections (which would be indicative of strong magnetic inhomogeneity) was observed on this temperature interval, it can be assumed that a noncollinear (canted) magnetic structure is involved in the transition from antiferromagnetism to ferromagnetism. The possible existence of such structure in alloys with competing interactions was theoretically predicted in [12, 13] and experimentally observed in several binary (Fe–Ni [14], Ni–Mn [15]) and ternary $(Fe(Pd_{1-x}Pt_x)_3$ [11, 16], $Fe(Pd_{1-x}Au_x)_3$ [10], $Fe_xMn_{1-x}Pt_3$ [17]) alloys.

Only ferromagnetic reflections of the type {100} and {110} were observed at $T > 112$ K. These reflections vanished at $T \approx 155$ K, the Curie temperature determined from magnetic measurements. There is no coherent magnetic neutron scattering at temperatures above 155 K.

Theoretical neutron diffraction patterns indicate that $\mu_{Fe}-\mu_{Pt} = (1.1\pm0.2)\mu_B$. An estimate of the magnetic moment per iron atom (under the assumption that the magnetic moments on the platinum atoms are close to zero) from the magnetization yielded $\mu_{Fe} = 0.85\mu_B$ (see Section 1.3), which is close to the difference between the magnetic moments of iron and platinum determined by neutron diffraction. These data are indicative of the small size of the induced moment on the platinum atoms.

The low values of the magnetic moments on the iron atoms on this interval may be due either to the appearance of chaotic ferromagnetic structure [12, 13] or a change in the magnitude of the magnetic moments on the iron atoms. On the basis of data regarding the magnitude and temperature dependence of the effective fields on the iron nuclei [3, 4], we can conclude that the first possibility is what is valid for this case.

In confirmation of results on the differential magnetic permeability and magnetization in weak magnetic fields, only the magnetic reflections corresponding to a structure with wave vector $\mathbf{k} = [1/2\ 0\ 0]$ are observed in the alloy with 30.5 at% Fe. Coherent magnetic scattering vanishes at a temperature $\sim 125\,\mathrm{K}$, which is similar to the magnetic ordering temperature values obtained from measurements of other properties.

Alloys with $c_{\mathrm{Fe}} = 28\,\mathrm{at}\%$ Our research results confirm those of Bacon and Men'shikov: only the reflections corresponding to the antiferromagnetic structure with wave vector $\mathbf{k} = [1/2\ 1/2\ 0]$ characterized by the Lifshits star $\{k_{11}\}$ in the simple cubical lattice Γ_c are observed for $T_1 < T < T_2$. The intensity of the $\{1/2\ 1/2\ 0\}$ and $\{1/2\ 1/2\ 1\}$-type reflections drops off precipitously at $T = T_1$, and the $\{0\ 0\ 1/2\}$, $\{0\ 0\ 3/2\}$, and $\{1\ 1\ 1/2\}$-type reflections corresponding to an antiferromagnetic structure with $\mathbf{k} = [1/2\ 0\ 0]$ begin to appear. The temperatures of the phase transitions determined by neutron diffraction coincide with the temperatures T_1 and T_2 obtained from magnetic, electrical, and other measurements.

The equivalent planes in this monocrystal also yield reflections of identical intensity, which indicates that a multi-domain structure evidently exists. It is, however, more difficult to determine the path along the $\{k_{11}\}$ star, since the structure factors for single-beam and multiple-beam paths turn out to be identical when multiple domains are taken into account. Sidorov and co-workers [11] analyzed the experimental data obtained in [18], and showed that the ternary isostructural alloy $Fe(Pd_{0.53}Pt_{0.47})_3$ has a single-beam path along the star $\{k_{11}\}$. We can therefore assume that the same case obtains for Pt–Fe alloys. This conclusion is supported by the existence of anisotropic spin wave dispersion in Pt_3Fe [19] resulting from the tetragonal symmetry of the magnetic lattice.

The simultaneous existence of reflections of both types ($\mathbf{k} = [1/2\ 1/2\ 0]$ and $[0\ 0\ 1/2]$) indicates that the two magnetic structures are, in Bacon's opinion [1], evidence that the two magnetic structures coexist. A second idea, which seems preferable to us, is that there is a noncollinear antiferromagnetic structure where the magnetic moments of the iron atoms deviate from the $< 001 >$ axis by some angle ϕ, and completely compensate for one another within an elementary cell. An example of this type of structure is shown in Fig. 5 (see inset). In this case, the projections of the moments on the Z-axis yield reflections typical of structures

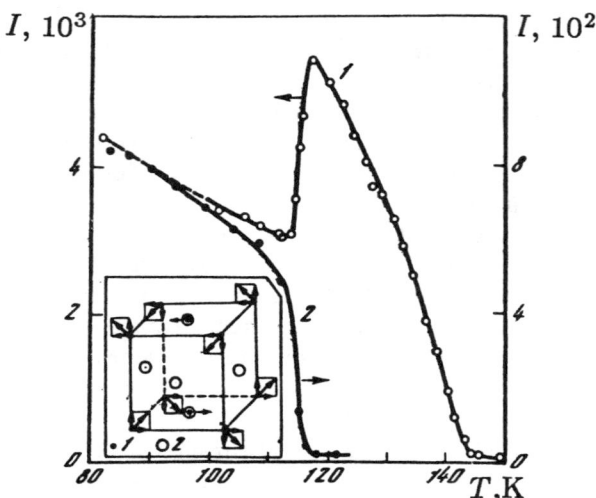

Figure 5   Elastic neutron scattering off various crystallographic planes in the $Pt_{72}Fe_{28}$ as a function of temperature. (1) (1/2 1/2 0); (2) (1 1 1/2). Inset: Possible type of magnetic structure for Pt–Fe alloys with Fe concentration ranging from 25.2 to 30 at% for $T < T_1$ and $\phi = 45°$: (1) Fe; (2) Pt.

with $\mathbf{k} = [1/2\ 1/2\ 0]$, while the projections on the X-axis yield reflections typical of $\mathbf{k} = [0\ 0\ 1/2]$. The relative intensities of these reflections are determined by the angle ϕ.

1.3. DETERMINATION OF MAGNETIC STRUCTURES FROM MAGNETIC MEASUREMENTS

The strong magnetic fields (with strengths of up to 150 kOe) used were produced using the "Solenoid" facility at the Institute for General Physics of the USSR Academy of Sciences, while the weak magnetic fields (up to 1 kOe) were obtained using a solenoidal coil with a 0.29 mm-diameter water-resistant enamel wire winding. The magnetization was measured either ballistically using an F 190 photocompensating microwebermeter with a sensitivity of 2 μWb/div or by using a high-oscillation-amplitude ($\approx$ 2.6 mm) magnetometer. These methods were chosen because of the inadequate sensitivity and high noise in the "Solenoid" facility, which rendered the use of ordinary vibrating magnetometers impossible.

In the ballistic measurements, a special mechanical device which converted the rotational motion of an electric motor axle into the reciprocal motion of a thin-walled stainless-steel tube (to which the sample holder was attached) slowly moved the sample into and out of a compensated measuring coil. The microwebermeter recorded the change in magnetic flux passing through the coil as the sample was moved into it. The coil was calibrated using nickel reference samples of the same size and shape as the samples being studied.

The relative error in the measurements of the magnetization as a function of temperature and magnetic field strength was no greater than 2.5%, while the absolute error was no greater than 8%. The sensitivity was $10^{-3}\,\mathrm{G}\cdot\mathrm{cm}^3$.

The sample was moved back and forth at a frequency of 25 Hz by a variable-current synchronous motor. The signal was recorded using compensated measuring coils.

The magnetometer electronics consisted of the following standard pieces of equipment: a U2-8 discriminator/amplifier and a V9-2 synchronous detector. The synchronous detector reference signal came from a sealed magnetic switch controlled by a permanent magnet oscillating at the magnetometer operating frequency.

Magnetization and Inverse Magnetic Susceptibility   The specific magnetization σ ($\sigma = J/D$, where J is the magnetization per unit volume and D is the density of the sample) measured as a function of temperature in an 11-kOe field for samples of Pt–Fe alloys having various compositions (see Fig. 2b in [8]) indicate that there are pronounced anomalies in magnetization near the first-order phase transition temperature T_1. The absence of a significant anomaly in σ at the Neel point is noteworthy. This is due to the fact that the peak at T_N is suppressed in weak magnetic fields.

Measurements of the inverse magnetic susceptibility $\chi^{-1} = (J/H)^{-1}$ as a function of temperature in a 20-kOe field (see Fig. 4 in [8]) showed that the function $\chi^{-1}(T)$ was linear at high temperature, i. e., that the Curie-Weiss law was satisfied at paramagnetic temperatures for all the alloys.

All of the alloys have the common characteristic that the functions $\chi^{-1}(T)$ deviate from linearity for $T < T_1$. The alloy with iron concentration 25.2 at% was also observed to have deviations from linearity in $\chi^{-1}(T)$, while no anomalies were observed for the alloy with $c_{\mathrm{Fe}} = 28$ at%. On the temperature interval $T_1 < T < T_2$, the derivative $\partial\chi/\partial T < 0$ for the alloys with $c_{\mathrm{Fe}} < 30.5$ at%, which leads us to believe that "ideal" antiferromagnetic ordering does not exist in this temperature region.

Extrapolation of the linear portions of the $\chi^{-1}(T)$ curves to the temperature axis yields the paramagnetic Curie temperature Θ_p; the slope can be used to calculate the Curie-Weiss constants C_{CW} and the effective magnetic moments μ_{eff} per formula unit $\mathrm{Fe}_c\mathrm{Pt}_{1-c}$. These data are presented in Table 2, which also includes the effective magnetic moments per iron atom and per "additional" iron atom (in excess of 25 at%). The magnetic moment per additional Fe atom was calculated from the function $(\Delta\chi)^{-1} = f(T)$, where $\Delta\chi = \chi_{\mathrm{alloy}} - \chi_{\mathrm{lattice}}$. The lattice susceptibility χ_{lattice} was assumed to be the same as that of the alloy with Fe concentration 24.8 at%.

Increasing iron concentration initially leads to a decrease in the absolute value of Θ_p, and then to a change in sign from negative to positive, which is indicative of the increasing contribution from ferromagnetic exchange interactions to the establishment of magnetic order. Similar results were obtained in [20].

The data presented in Table 2 imply that the magnetic moments per iron atom determined in the paramagnetic temperature region (with a g-factor of 2, $\mu_{\text{Fe}} = (4.3 \pm 0.2)\mu_B$ for alloys with $c_{\text{Fe}} < 30$ at%) are larger than those determined at lower temperatures from the neutron diffraction data ($\mu_{\text{Fe}} = 3.3\mu_B$ [1]). A similar result was obtained in [21] for alloys with iron concentration 22.6 at%. This was because of the fact that both the iron and the platinum atoms contribute to the paramagnetic susceptibility. The higher values of the effective moments for the "additional" iron atoms is clear evidence that the $5d$ electrons in the platinum atoms are polarized by the nearest-neighbor iron atoms.

Further increases in the effective moment per formula unit $\text{Fe}_c\text{Pt}_{1-c}$ in the alloys with iron concentration greater than 30 at% indicate that the "polarization clouds" are beginning to overlap and form what amount to complexes or clusters that extend over at least several elementary cells and are in a superparamagnetic state at high temperatures.

The superparamagnetic behavior of the alloys at high temperatures is confirmed by the function describing magnetization as a function of magnetic field strength in strong magnetic fields (see Fig. 10 in [8]). The magnetization isotherms are

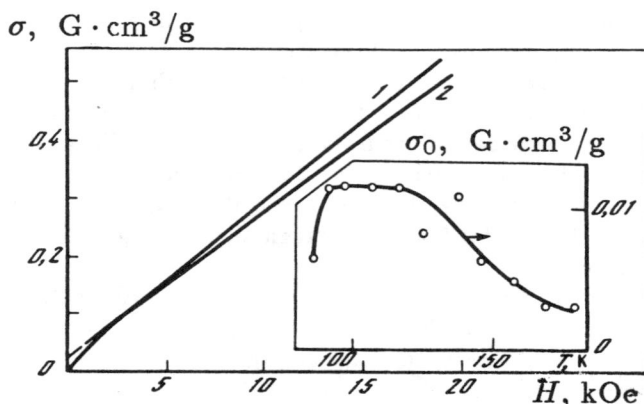

Figure 6   Magnetization of the 25.2 at% Fe alloy as a function of magnetic field strength. (1) $T = 88$ K; (2) $T = 93$ K. Inset: "Spontaneous" magnetization of this alloy as a function of temperature.

Table 2. Parameters Determined From the Curie–Weiss Law for Pt–Fe Alloys

c_{Fe}, at%	Θ_p, K	μ_{eff}/Fe_cPt_{1-c}, μ_B	μ_{eff}/Fe atom, μ_B	$\mu_{eff}/addit.$ Fe atom, μ_B
24.8	-90	2.6	5.3	—
25.2	-50	2.6	5.3	52±10
27.0	20	2.8	5.4	10
28.0	40	2.8	5.2	9
30.5	165	4.6	8.1	19
32.2	190	4.6	8.0	17

NOTE: Observational errors: ±10 K for Θ_p, ±0.2μ_B for μ_{eff}/Fe_cPt_{1-c} and μ_{eff}/Fe atom, ±1μ_B for $\mu_{eff}/addit.$ Fe atom.

nonlinear, while the susceptibility is fairly high ($\chi \sim 10^{-3}$ for alloys with $c_{Fe} \geq$ 30.5 at%).

Magnetic Properties of the Alloys on the Temperature Interval $T_1 < T < T_2$ The Pt–Fe alloys have nonlinear magnetization as a function of magnetic field strength at high magnetic fields on this temperature interval. Deviations from linearity also appear for small fields ($H = 2$–3 kOe) in the alloys that are between 25.2 and 28 at% Fe. The magnetization as a function of magnetic field strength for a monocrystal with 25.2 at% Fe, measured at $T = 88$ K and 93 K for fields of up to 18.5 kOe using a vibrating magnetometer are shown in Fig. 6. The function $\sigma(H)$ has the characteristic form for antiferromagnetic materials with a small ferromagnetic component. The temperature dependence of the magnetization σ_0 determined by extrapolating the linear part of $\sigma(H)$ from the region $H > 3$ kOe back to $H = 0$ is shown in the right-hand part of Fig. 6. Approximately the same value, $\sigma_0 = 0.01$–0.02 G · cm^3/g, was also obtained for the 27 at% Fe alloy.

The fact that the quantity σ_0 is a function of temperature indicates that there is either a coherent or incoherent canted structure with a weak, non-zero ferromagnetic component having a magnetization equal to approximately 0.04% that of disordered ferromagnetic alloys of the same composition. This corresponds to a very small angular deviation in the magnetic moments ($\phi \approx 2'$); this is why no ferromagnetic reflections were observed for $T_1 < T < T_2$.

In the alloy with Fe concentration 30.5 at%, the magnetization also peaks in the same temperature region where the differential magnetic susceptibility is observed to have a peak (Fig. 7a).

The weak-field magnetization isotherms for alloys with $c_{Fe} > 30.5$ at% (Fig. 7b) imply that these alloys are ferromagnetic between T_1 and T_2, even though measurements in strong fields (see Fig. 10 in [8]) indicate that they do not become fully saturated, even at fields as high as 150 kOe. The spontaneous magnetiza-

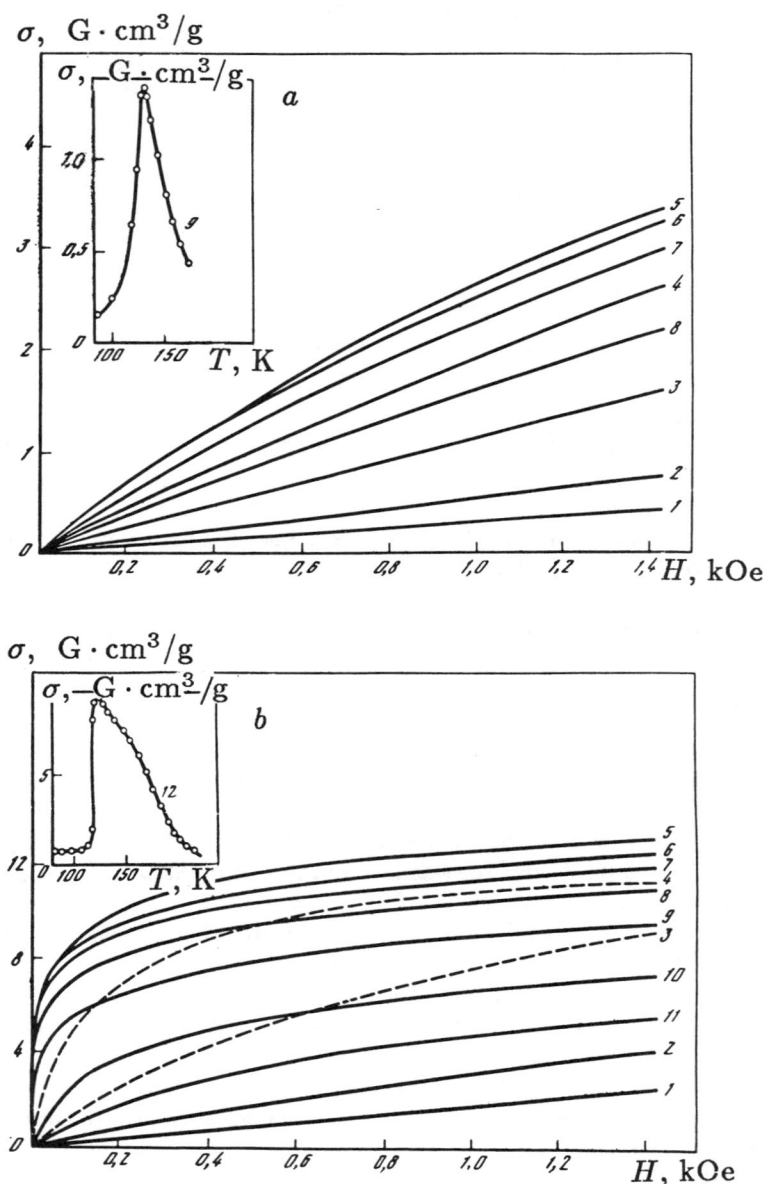

Figure 7   Magnetization as a function of magnetic field for iron–platinum alloys with iron concentrations of 30.5 (a) and 32.2 at% (b) for small fields and various T, K. a) (1) 77; (2) 102.5; (3) 117; (4) 123; (5) 128; (6) 134; (7) 140; (8) 156 K; (9) $\sigma(T)$ in a 470-Oe field; b) (1) 84; (2) 93.5; (3) 105; (4) 111; (5) 115.5; (6) 126; (7) 137.5; (8) 145; (9) 153; (10) 163; (11) 172 K; (12) $\sigma(T)$ in a 470-Oe field.

tion σ_s and Curie temperature can be determined by constructing $\sigma^2 = f(H/\sigma)$ for weak magnetic fields. The values obtained for σ_s turned out to be 2–2.5 times smaller than for disordered Pt–Fe alloys at the same value of T/T_C. The value $\sigma_s = 10\,\mathrm{G}\cdot\mathrm{cm}^3/\mathrm{g}$ implies that $\mu/\mathrm{Fe}_c\mathrm{Pt}_{1-c} = 0.3\mu_B$ or $\mu/\mathrm{atFe} = 0.85\mu_B$. Because the functions $\sigma^2 = f(H/\sigma)$ are nonlinear, extrapolation from the strong-field region ($H \geq 90\,\mathrm{kOe}$) yields large values of σ_s; however, even these values are a factor of 1.5–2 smaller than the values of σ_s for disordered alloys.

Hysteresis effects (coercive force $H_c \approx 3\,\mathrm{Oe}$) are observed in the ferromagnetic ordering region. There is no anomaly at $T = T_3$ in the measurements of the magnetization for constant fields. The residual magnetization as a function of temperature is also evidence for the presence of ferromagnetism in alloys with $c_{\mathrm{Fe}} > 30.5\,\mathrm{at\%}$ on the temperature interval T_1–T_2.

The alloys with iron concentrations of 34 and 36 at% are ferromagnetic all the

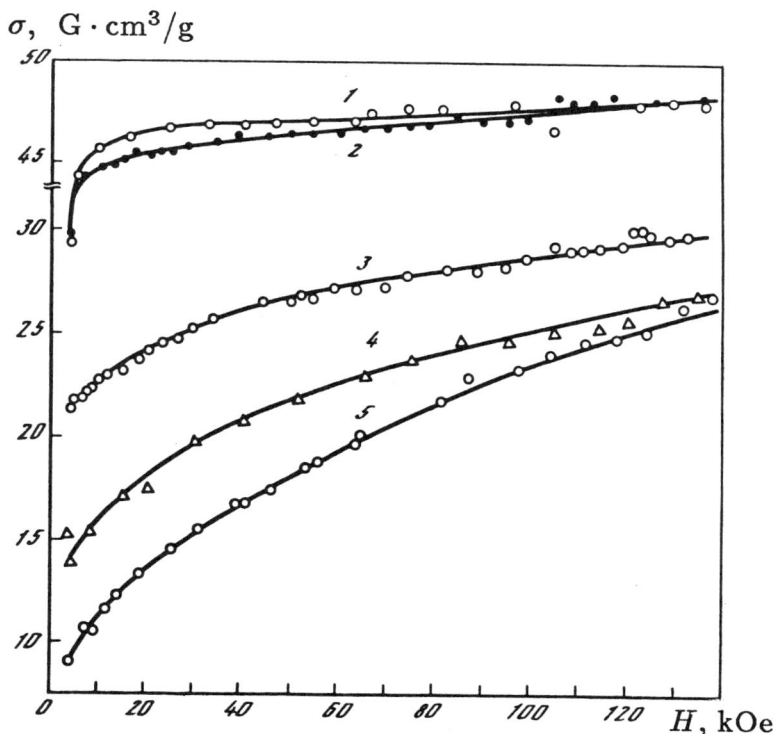

Figure 8   Magnetization isotherms for a polycrystalline iron–platinum alloy with 36 at% Fe. T, K: (1) 4; (2) 77; (3) 291; (4) 336; (5) 357.

way up to their Curie temperatures. The residual magnetization and magnetization measured in weak magnetic fields decrease sharply near T_C, while the
differential magnetic susceptibility reaches a maximum at this same point (see
Fig. 2). The magnetization isotherms measured for the alloys with $c_{Fe} = 36$ at%
in strong magnetic fields are shown in Fig. 8. It should be noted that the functions $\sigma^2 = f(H/\sigma)$ are also nonlinear for these alloys; the transition is spread out
somewhat at the Curie point, and superparamagnetic behavior is observed above
T_C.

1.4. PHASE TRANSITIONS

Let us discuss the phase transitions at T_1 and T_2. As was already noted above,
the phase transitions at $T = T_1$ (line BO in Fig. 3. for alloys with $c_{Fe} < 30.5$ at%
are accompanied by various anomalies in the magnetic, electrical, and galvanometric properties characteristic of first-order phase transitions: "jump discontinuities"
in the physical properties and temperature hysteresis. The transition at $T = T_2$
(line AO) is a second-order phase transition.

However, the phase transitions in alloys with iron concentration greater than
30.5 at% are not typical second-order phase transitions: There are no sharp changes
in either σ or ρ (lines OD and OC in the diagram). The presence of anomalies in the heat capacity at $T = T_1$ and T_2 in the $c_{Fe} = 28$ at% alloy and the
lack of anomalies in the alloy with iron concentration 32 at% is additional confirmation of the fact that the phase transitions for iron concentrations greater
than 32 at% Fe (Fig. 9) are different in nature.[1]

The fact that the phase transitions are indeed different in nature is also confirmed by the sharpness of the $\mathbf{k} = [1/2\ 0\ 0]$ and $[1/2\ 1/2\ 0]$ antiferromagnetic
reflections in the alloy with 28 at% Fe and the gradual, "drawn out way" that the
reflections fall off in the alloy with $c_{Fe} = 32$ at%. The transitions in alloys with
$c_{Fe} > 30$ at% are apparently second-order phase transitions, but they are washed
out due to the strong magnetic interactions.

The change in entropy in the first-order phase transition at $T = T_1$ can be
calculated from the formula

$$\Delta S_1 = \int_{T_1 - \Delta T}^{T_1 + \Delta T} c_p(T)/T\, dT. \tag{1}$$

The calculation yields $\Delta S_1 = 140\,\text{mJ} \cdot \text{mole}^{-1} \cdot \text{K}^{-1} = 0.9\,\text{mJ} \cdot \text{g}^{-1}\text{K}^{-1}$. This
quantity is more than an order of magnitude smaller than the change in entropy
during the transition from the antiferromagnetic to ferromagnetic states in the

---

[1]The heat capacities of these alloys were measured as a function of temperature from 4.2
to 300 K on V. M. Polovov's apparatus at the Institute for Solid State Physics (USSR Academy
of Sciences).

C_p, J/(g · atom)

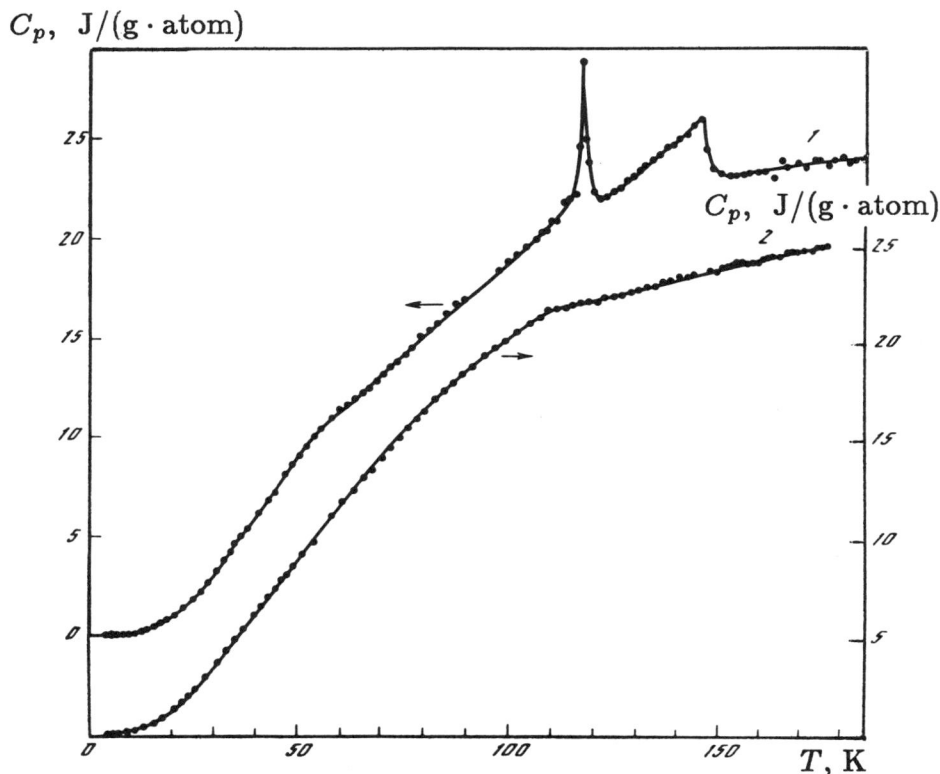

Figure 9   Heat capacities for iron–platinum alloys as a function of temperature. (1) 28 at% Fe; (2) 32 at% Fe.

equiatomic Fe–Rh alloy ($\Delta S = 5$–$19\,\mathrm{mJ} \cdot \mathrm{g}^{-1} \cdot \mathrm{K}^{-1}$ [22]). The "lattice" contribution to ΔS,

$$\Delta S_{\mathrm{lattice}} = 3\alpha\Delta V/V(\kappa D)^{-1}, \qquad (2)$$

(where α is the coefficient of thermal expansion, κ the compressibility, D the density, and $\Delta V/V$ the relative change in volume) is quite small: $\Delta S_{\mathrm{lat}} \approx 5 \cdot 10^{-5}\,\mathrm{J} \cdot \mathrm{g}^{-1} \cdot \mathrm{K}^{-1}$ for $\Delta V/V \approx 10^{-4}$ [23].

If we assume that the change in entropy in the transition at T_1 is entirely a result of the change in the electron part of the entropy, we then have

$$\Delta S_1 = (\gamma_1 - \gamma_2)T_1 = \Delta\gamma T_1 \tag{3}$$

whence we find that $\Delta\gamma = 1.2\,\mathrm{mJ \cdot mole^{-1}K^{-2}}$, which is, to order of magnitude, in agreement with the measured electron heat capacity of the $c_{Fe} = 28\,\mathrm{at\%}$ alloy determined at low temperatures ($\gamma = 2.7\,\mathrm{mJ \cdot mole^{-1}K^{-2}}$). This means that we can quite possibly have changes in the magnetic part of the entropy can quite possibly be accompanied by changes in the electronic part of the entropy.

The change in entropy during the transition through the Neel point can be determined from the expression

$$\Delta S_2 = \int_0^{T_2} C_m(T)/T\,dT, \tag{4}$$

$$C_m(T) = C_p(T) - [C_{el}(T) + C_D(T)],$$

where $C_m(T)$, $C_{el}(T)$, and $C_D(T)$ are the "magnetic," "electron," and "Debye" (lattice) parts of the heat capacity, respectively.

It is not possible to completely rigorously separate the magnetic part of the heat capacity from the measured total heat capacity, since the the Debye temperature is a function of temperature in the present case. Moreover, since T_2 and Θ_D are of the same order of magnitude, we must introduce a correction for anharmonicity.

However, a qualitative estimate of the change in entropy using the formula

$$\Delta S_2 = \int_{T_1}^{T_2} [C_1(T) - C^*(T)]/T\,dT \tag{6}$$

($C^*(T)$ is a smooth function obtained by extrapolating $C_p(T)$ from the region $T > T_2$ to $C_p(T)$ in the region $T < T_1$) should yield the correct order of magnitude. Our data indicate that $S_2 \approx 216\,\mathrm{mJ \cdot mole^{-1} \cdot K^{-1}} = 1.4\,\mathrm{mJ \cdot g^{-1}K^{-1}}$.

For localized spins, $\Delta S_2 = R\ln(2S + 1)$, where R is the gas constant and S is the spin, i. e., $\Delta S_2 = 5.75\,\mathrm{mJ \cdot mole^{-1} \cdot K^{-1}}$, which is at least an order of magnitude greater than the value obtained for the Pt–Fe alloy. This implies that, like chromium ($\Delta S_2 = 18\,\mathrm{mJ \cdot g \cdot at^{-1} \cdot mole^{-1} \cdot K^{-1}}$ [24]), CrB_2 ($\Delta S = 1\,\mathrm{mJ \cdot mole^{-1} \cdot K^{-1}}$ [25]), and several other substances, the alloy with $28\,\mathrm{at\%}$ Fe has properties characteristic of zonal magnetic substances.

1.5. COMPOSITION-INDUCED PHASE TRANSITIONS FROM ANTIFERROMAGNETISM TO FERROMAGNETISM

In recent years, a great deal of attention has been devoted to experimental and theoretical research on systems in which composition-induced phase transitions from antiferromagnetism to ferromagnetism occur within the same crystallographic modification. These phase transitions have been observed in binary alloys such as Ni–Mn [26], Fe–Cr [27, 28], and Pt–Mn [29]; disordered ternary alloys such as Fe–Ni–Mn [30] and Fe–Ni–Cr [31]; and ordered ternary alloys such as $Fe(Pd_xPt_{1-x})_3$ [32], $Fe(Pd_xAu_{1-x})_3$ [10], $(Fe_xMn_{1-x})Pt_3$ [17], and $(Fe_{1-x}Mn_x)Pt$ [33], as well as in semiconductors such as, for example, $Cd_{1-x}Zn_xCr_2Se_4$ and $Hg_{1-x}Zn_xCr_2Se_4$ [34] and systems containing metals and nonmetals, such as $Co(S_xSe_{1-x})_2$ [35], etc.

The most complete theoretical studies of magnetic phase diagrams for systems with chaotic exchange coupling and various "competing" exchange interactions within the framework of the localized magnetic moment model have been published by Medvedev and coworkers [12, 13, 36, 37]. The calculations were carried out for completely disordered Ising and Heisenberg magnetic substances that have formed crystals with simple cubic, bcc, and fcc structures.

It was shown that in the simple and bcc lattices, phase transitions out of the paramagnetic state can occur from the following three states as the temperature decreases: a collinear chaotic ferromagnet, a collinear chaotic antiferromagnet, or a state with a coplanar canted structure that is a superposition of a ferromagnet along one coordinate axis and an antiferromagnet along another (the term "chaotic structure" means that the majority of the magnetic moments are aligned parallel to the direction of spontaneous magnetization, while a minority are aligned antiparallel to this direction). As the temperature decreases further, the system undergoes a transition into one of the states with spin projections "frozen" perpendicular to the magnetization: asperomagnetic, antiasperomagnetic, and noncoplanar canted states.

On the other hand, the spin glass state is a fairly typical example of a magnetic substance with an fcc lattice. Moreover, the magnetic structures predicted for the simple and bcc lattices can also exist in magnetic substances of this type.

The experimental results obtained for the crystalline structures discussed above generally support the theoretical results [10, 17, 27, 30, 31].

As it turns out, the localized magnetic moment model can also be used in describing the magnetic structure of transition metal alloys, since the spin density in such alloys is essentially localized around the crystalline lattice points. The neutron diffraction experiments of Kohgi and Ishikawa [19, 38] confirmed the applicability of the Heisenberg model to the stoichiometric alloy Pt_3Fe. We may therefore assume that the complex magnetic structures formed in ordered Pt–Fe alloys with compositions in the transition region between antiferromagnetism and ferromagnetism will be analogous to those predicted by Medvedev and coworkers.

Iron-platinum alloys also possess several characteristics that lead to a wider variety of behavior than in the model systems.

We shall now list the types of large-scale magnetic ordering which occur in ordered Pt–Fe alloys.

The stoichiometric-composition alloy Pt_3Fe An antiferromagnetic structure which can be described by a single-beam transmission path along the Lifshits star $\{k_{11}\}$ in a simple cubic lattice appears at $T < T_N = 170\,\text{K}$, and this structure is maintained down to low temperatures.

Alloys with $25.2 \leq c_{Fe} \leq 28\,\text{at}\%$ As the temperature decreases below T_2, an antiferromagnetic structure with $k = [1/2\ 1/2\ 0]$ is formed; however, this structure will have distortions introduced by the additional iron atoms and possibly by the polarized platinum atoms. According to theory [12, 13], a collinear, but chaotic, antiferromagnetic structure will form just below the magnetic ordering temperature, while transverse projections of the magnetic moments appear at lower temperatures. The temperature dependence of the "spontaneous moment" σ_0 (see Fig. 6) may in fact indicate that such a transition might occur.

If a coherent canted magnetic structure (which is conducive to the formation of tetragonal distortions in the crystal) is formed in the process, the deviations from linearity we observed in the magnetization isotherms for both strong and weak fields become understandable.

However, the results in [12, 13, 36, 37] indicate that a coherent canted structure may in fact not exist for low ferromagnetic bond densities. The distortions in the antiferromagnetic structure will then be local in nature, so that the nonlinearities in the $\sigma(H)$ curves might be due to the ordering of individual clusters in such regions of the magnetic field.

As the temperature decreases further, the magnetic moments become reoriented at $T = T_1$, with the formation of noncollinear antiferromagnetic structure characterized by the two Lifshits stars $\{k_{11}\}$ and $\{k_{10}\}$. The distortions in the crystalline lattice observed by Bacon and Wilson [23] and the associated changes in the zonal structure and crystallographic anisotropy constant are evidently characteristic of this phase transition.

Alloys with $30.5 \leq c_{Fe} \leq 32.2\,\text{at}\%$ The alloys with $c_{Fe} \geq 30.5\,\text{at}\%$ have $\Theta_p > 0$, i. e., the positive exchange interactions predominate over the negative. The increase in $\mu_{eff}/Fe_c Pt_{1-c}$ and the substantial nonlinearity in the $\sigma(H)$ curves for $T > T_2$ indicate that the clusters that are formed already include several elementary cells and at high temperatures, they behave like superparamagnetic particles.

According to [12, 13, 36], as the temperature decreases to $T = T_2$ chaotic ferromagnetic structure in which most of the moments are aligned parallel to the direction of magnetization and a minority are antiparallel (which is what in fact gives rise to the less-than-nominal value observed for the spontaneous magnetization) can form in alloys with $c_{Fe} > 30.5\,\text{at}\%$. Direct positive interaction between nearest-neighbor iron atoms may play an important role in the formation of this structure, since the onset of ferromagnetic ordering occurs at compositions for which the presence of two additional iron atoms in a single elementary cell becomes highly probable (at $c_{Fe} = 30$ at%, there is one additional iron atom for every five elementary cells, at $c_{Fe} = 30$ at%, one for every four cells, and at $c_{Fe} = 30$ at%, one for every three cells).

As the temperature decreases further, a noncollinear magnetic structure (possibly the coplanar canted structure predicted by Medvedev [12, 13]) is formed between T_1 and T_2. One might imagine that the magnetic moments rotate more or less smoothly from parallel to antiparallel over the region in which this structure exists. This temperature interval is evidently very narrow for the alloys with approximately 30.65 at% Fe, and only a single anomaly appears in the physical properties.

The clusters are not free on the temperature interval T_1–T_2, since they are in the nonzero molecular field of the ferromagnetic structure. For $T < T_1$, the magnetic structure consists of an antiferromagnetic matrix formed by the atoms in the main iron lattice and described by the wave vector $\mathbf{k} = [1/2\ 0\ 0]$ in which the clusters are embedded (see Section 2).

The "critical alloy" with 30.5 at% Fe deserves special attention in this discussion. The density of ferromagnetic bonds in this alloy is still not large enough for the onset of ferromagnetic ordering at $T > T_1$, but clusters with large moments do exist, and a transition from the paramagnetic state to the antiferromagnetic state characterized by the wave vector $\mathbf{k} = [1/2\ 0\ 0]$ does occur with the clusters embedded in the alloy.

Alloys with $c_{Fe} = 34$–36 at% In these alloys, essentially all of the elementary cells have one or more "additional" iron atoms. Because of the extreme magnetic inhomogeneity, the transition to ferromagnetic order is is fairly "washed out." Since there are negative interactions as well as positive interactions in the alloys, a transition to a canted structure with antiferromagnetic reflections may occur as the temperature decreases.

The noncollinearity may be overcome fairly easily by external magnetic fields: the magnetization saturates even for fields $H \geq 10$ kOe. The fact that the magnetization of the ordered alloy is lower than that of the disordered alloy (35 and 43 G $\cdot$ cm^3/g, respectively) for alloys with $c_{Fe} = 34$ at%, while the magnetization is nearly the same in the ordered and disordered states ($\sigma_s \approx 46$ G $\cdot$ cm^3/g at $T = 4.2$ K) of the alloy containing 36 at% Fe is quite interesting. This is either because of the fact that chaotic ferromagnetic structure is still present in the alloy with $c_{Fe} = 34$ at% or because of the fact that the moments on the iron or platinum atoms are lower in the ordered alloy than in the disordered alloy because of the differences between the zonal structures of the alloys.

And so, our research on the physical properties of Pt–Fe alloys led to the first determination of the magnetic c–T phase diagram for alloys with compositions in the transition region between antiferromagnetism and ferromagnetism. The types of large-scale magnetic order in the alloys and the properties of the phase transitions were determined, and it was shown that the localized magnetic moment model can be used to describe the magnetic structures formed to first approximation.

2.   CHAOTIC MAGNETIC STRUCTURES IN Pt–Fe AND Fe–Rh ALLOYS

As was noted above, the theoretical work of Medvedev and coworkers [12] implies that one might expect the formation of various chaotic magnetic states in alloys of these two systems for compositions in the transition region from antiferromagnetism to ferromagnetism. We shall now discuss this problem in some detail.

2.1.   EXPERIMENTAL RESULTS

The research carried out here indicated that at low temperatures, large-scale magnetic order is accompanied by local magnetic order in alloys of both systems when the compositions differ from stoichiometric. The formation of this local order mainly makes itself felt in the residual magnetization σ_r for antiferromagnetic alloys and in the fact that it increases rapidly rather than proportional to composition (i. e., with increasing iron concentration). The values of σ_r at 4.2 K after magnetization in fields of up to 130 kOe are given for Pt–Fe and Fe–Rh alloys of various composition in Table 3.

The residual magnetization in Pt–Fe alloys increases sharply for $c_{Fe} \geq 30$ at%, and is also large in ferromagnetic alloys. One interesting characteristic of the alloys with $c_{Fe} \geq 30.5$ at% is that the residual magnetization is anisotropic. As can be seen from Table 3, the maximum value of σ_r occurs along the $< 110 >$ axis.

In Fe–Rh alloys, the residual magnetization is largest for compositions in the transition region from antiferromagnetism to ferromagnetism at $c_{Fe} = 51.0$ at%. In Fe–Rh alloys, however, σ_r is quite sensitive to the shape of the sample because of the effects of internal degaussing fields. Table 3 includes data for samples with various degaussing factors: $l/d_{eff} \approx 5$–7 ($N \sim 0.5$) and $l/d_{eff} \approx 1.7$–1.8 ($N \sim 2.5$), where $d_{eff} = 4S/\pi)^{1/2}$, l is the length, and S is the cross-sectional area of the sample.

The alloys can be divided arbitrarily into three groups differing in iron concentration with respect to the physical properties we have determined:

Alloys with Small Deviations from Stoichiometric Composition
Fig. 10 shows the residual magnetizations of Pt–Fe and Fe–Rh alloys of various composition as a function of temperature. The residual magnetization is small for

Table 3. Residual Magnetization in Pt–Fe and Fe–Rh Alloys at 4.2 K

c_{Fe}, at%	Magnetic Field Direction	σ_r, G·cm³/g	c_{Fe}, at%	σ_r, G·cm³/g	
				$N \sim 0.5$	$N \sim 2.5$
	Pt–Fe			Fe–Rh	
25.2	—	$2.5 \cdot 10^{-3}$	48.0	0.18	—
28.0	—	$7.2 \cdot 10^{-3}$	49.0	0.05	—
30.5	[110]	0.85	49.5	0.05	—
32.16	[111]	4.3	50.0	0.21	0.11
32.0 (cubical)	[110]	4.4	50.5	0.43	0.25
	[111]	4.3	50.75	1.6	0.90
	[001]	3.1	51.0	3.6	1.6
34.0	—	5.0	51.5	1.7	0.27
36.0	—	5.5	52.0	1.2	0.20

small deviations from stoichiometric composition in the Pt–Fe alloys; the magnitude of and variation in the residual magnetization as a function of temperature depend strongly on how the sample was treated beforehand. For example, after the samples were cooled to liquid-helium temperatures and measured in strong magnetic fields several times, residual stresses appeared in the samples, which led to an increase in the residual magnetization (Fig. 10d, Curves 3 and 4). Annealing at 600° for 6–8 hr removed these "induced defects" (Fig. 10d, Curves 1 and 2). It should be noted that at liquid-helium temperatures, a small amount of residual magnetization was even observed in samples of stoichiometric composition ($c_{Fe} = 24.8$–24.9 at%), although this magnetization rapidly vanished as the temperature was increased.

Similar results were also obtained for Fe–Rh alloys of equiatomic composition and small deviations from equiatomic composition.

Since these effects are also observed in carefully annealed alloys, it is natural to presume that they are at least partially due to the presence of "additional" iron atoms over and above the stoichiometric compositions Pt_3Fe and FeRh. Theoretical work indicates that small amounts of additional iron may lead to deviation of the magnetic moments from collinearity and the formation of canted structures (either coherent or noncoherent) with a weak ferromagnetic moment accompanied by the possible formation of small "inclusions" or clusters within the antiferromagnetic matrix.

The magnetization as a function of magnetic field strength at low temperature for the Pt–Fe alloys with 25.2 and 28 at% Fe shows small deviations from linearity for fields greater than 40–50 kOe (see Fig. 8 in [8]), while the functions $\sigma(H)$ are practically linear at higher temperatures. On this basis, we may presume that these alloys have individual clusters ordered by the field, with small resultant moments, rather than a canted structure. In Fe–Rh alloys, the functions $\sigma(H)$ at 4.2 K are practically linear in strong fields (Fig. 11).

Finally, partially spin-ordered states—the antiasperomagnetic state predicted in [12, 39] for systems with dominant antiferromagnetic interactions—may exist in these alloys at low temperatures.

Alloys with "Moderate" Deviations from Stoichiometric Composition The effects noted above become much stronger for larger deviations from stoichiometric composition: 5–7 at% additional iron atoms for Pt–Fe alloys and 0.75–1.0 at% for Fe–Rh alloys: σ_r sharply increases, and noticeable "time" effects appear. For example, σ_r in a $Pt_{67.84}Fe_{32.16}$ crystal decreased by 7.3% over a one-hour period after removal of the magnetic field. This effect was even more noticeable at 27 K: σ_r decreased by 14.5% over a period of 30 min and by 19% in 1.5 hr. This effect is less prominent for the alloy $Fe_{50.75}Rh_{49.25}$: At 4.2 K, σ_r decreased by only 3% over a 30 min period of time.

At low temperatures, the residual magnetization in Pt–Fe alloys as a function of temperature obeys an exponential law (Curve 2 in Fig. 10b and Curve 3 in Fig. 10c), which is a characteristic trait of cluster and spin glasses [40]. Deviations from this law are observed at $T = 20$–25 K.

This type of sharp increase in residual magnetization with decreasing temperature was not observed in the Fe–Rh alloys with $c_{Fe} = 50.75$ and 51.0%. Constructing $\log \sigma_r = f(T)$ (see inset in Fig. 10d), however, leads to the conclusion that (for the alloy with $c_{Fe} = 50.75$ at%) or $T \lesssim 35$ K (for the alloy with $c_{Fe} = 51.0$ at%) σ_r increases more rapidly for $T \lesssim 15$ K than for higher temperatures.

Shifted hysteresis loops are observed in Pt–Fe alloy samples when they are cooled in a magnetic field (Fig. 12), and the magnetization measured as a function of temperature upon cooling in zero and nonzero magnetic fields are quite different.

The shift in the hysteresis loops is much smaller for the iron–rhodium alloys, and is approximately 4–5 Oe in measurements at fields of up to ±240 Oe. The effects of cooling in a magnetic field are only apparent for measurements in weak magnetic fields (see inset in Fig. 13b).

These effects have also frequently been observed in spin and cluster glasses [41, 42]. These experimental facts imply that alloys containing between 30.5 and 32.2 at% Fe always have "frozen" clusters, i. e., there is small-scale magnetic order. Since these alloys also have large-scale magnetic order consisting of antiferromagnetic structure with wave vector $\mathbf{k} = [1/2\ 0\ 0]$; a more accurate name for this state would probably be "cluster glass embedded in an antiferromagnetic matrix."

The superparamagnetic behavior of these alloys observed at temperatures higher than the magnetic ordering temperature (see Section 1.3) also indicates that giant clusters are formed in place of a simple antiasperomagnetic state.

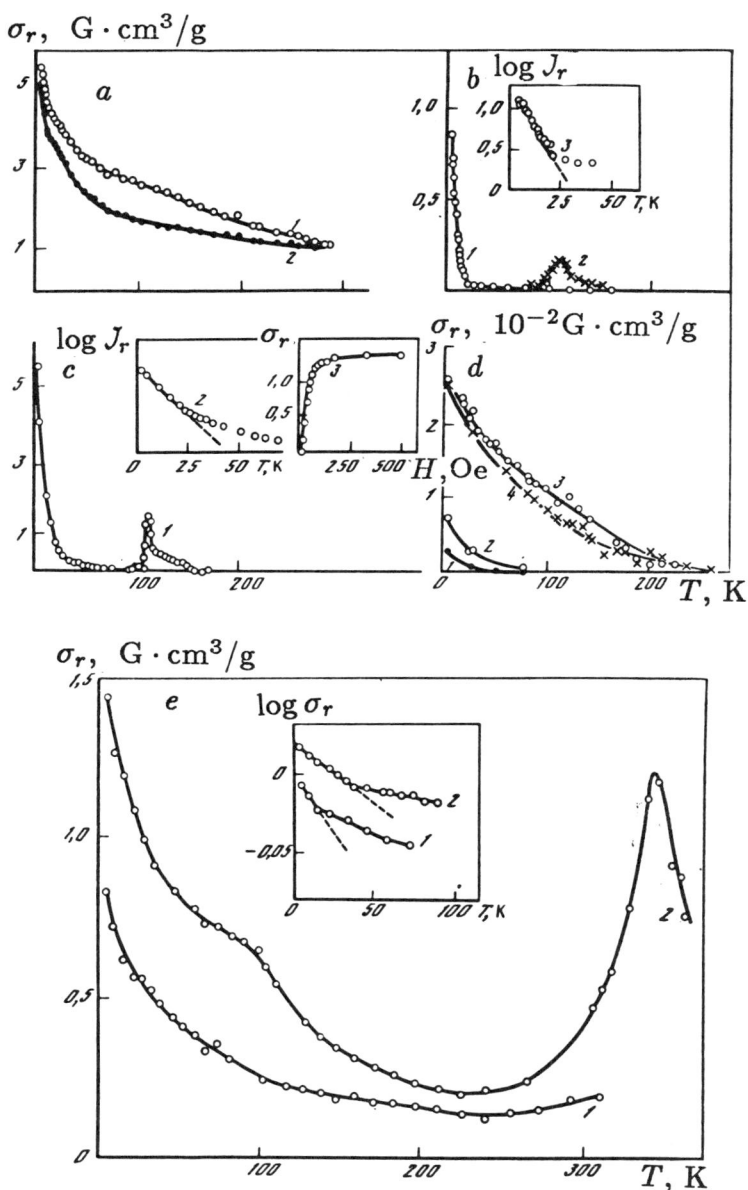

Figure 10   Residual magnetization as a function of temperature for iron–platinum (a-d) and iron–rhodium (e) alloys. (a) (1) 36; (2) 34 at% Fe; (b) 32.2 at% Fe ((1) $\sigma_r(T)$; (2) $\log J_r(T)$; (c) (1) 30.5 at% Fe; (2) 30.65 at% Fe; (3) $\log J_r(T)$; (d) (1), (3) 25.2 at% Fe; (2), (4) 28 at% Fe ((1) and (2) are measurements after annealing; (3) and (4) are measurements after numerous coolings to 4.2 K and measurements in magnetic fields of up to 150 kOe); (e) (1) 50.75 at% Fe; (2) 51.0 at% Fe (Inset: the function $\log \sigma_r(T)$).

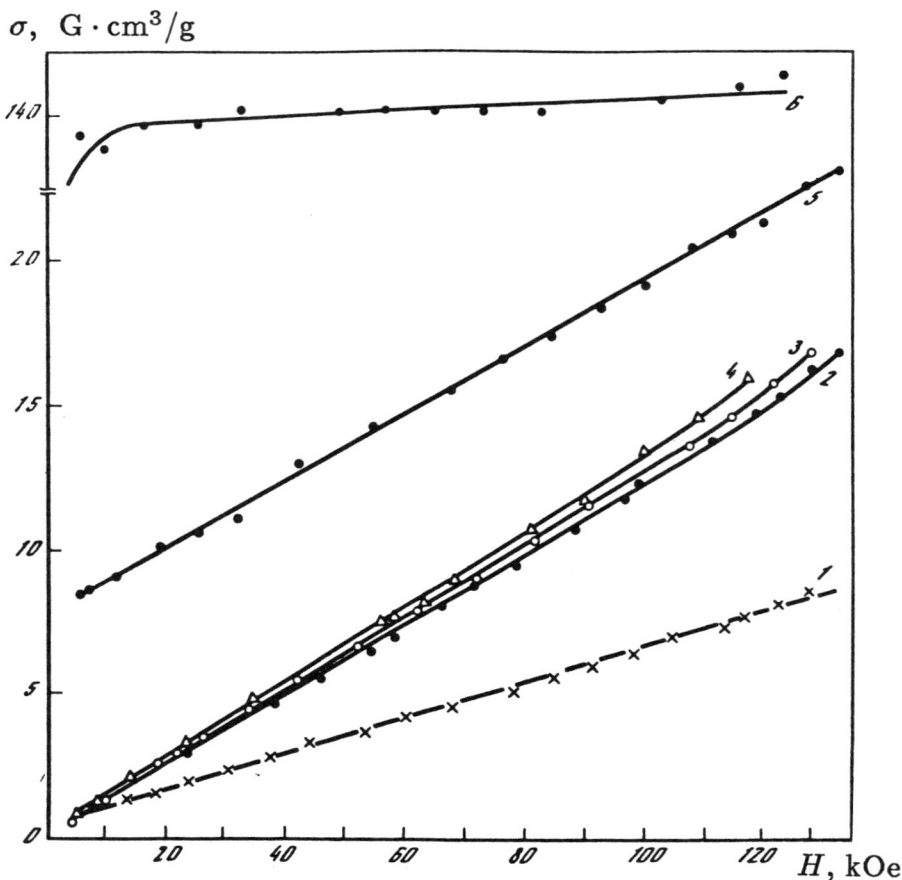

Figure 11   Specific magnetization of iron–rhodium alloys at 4.2 K as a function of field strength for strong magnetic fields.

The absence of neutron diffraction data at low temperatures for the 50.75 and 51.0 at% Fe iron–rhodium alloys means that we can only reach tentative conclusions on their magnetic state. These alloys are similar in behavior to the ferromagnetic alloys. For example, at liquid helium temperatures, the magnetization rapidly reaches saturation (Fig. 14), and hysteresis loops are observed when the material is remagnetized. Extrapolation of the magnetization to $H = 0$ allows us to calculate the "spontaneous magnetization," which is less than 5% of the magnetization in the ferromagnetic state. The ordering temperature T_o

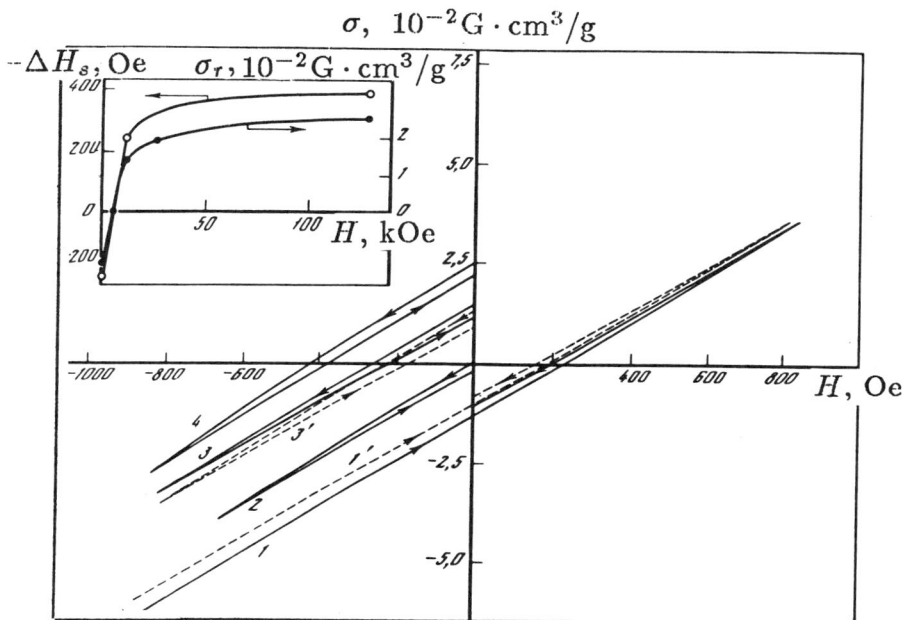

Figure 12   Shifted hysteresis loops at 4.2 K for a Pt–Fe alloy with $c_{Fe} = 28$ at%. (1, 1') After cooling in a field of -800 Oe field in the [001] and [110] directions, respectively; (2, 2') After exposure to a field $H = 5$ (2), $H = 12$ (3, 3'), and $H = 130$ kOe (4). Inset: size of shift ΔH_s in hysteresis loop and residual magnetization σ_r as a function of magnetic field.

determined from the Belov-Arrot construction $H/\sigma = f(\sigma^2)$ turned out to be equal to approximately 40 and 70 K for the alloys with $c_{Fe} = 50.75$ and 51.0 at%, respectively.

Since the strong-magnetic-field susceptibility of an equiatomic iron–rhodium alloy and alloys containing additional iron are approximately equal (see Fig. 11), and the magnetization increases by approximately the same amount in the transition to the ferromagnetic state when the temperature increases, one can naturally assume that large-scale antiferromagnetic order is also present in non-equiatomic Fe–Rh alloys at low temperatures.

So, the question is whether large-scale ferromagnetic order exists in these alloys, as it does in the Fe–Rh–Ir alloys [43], i. e., whether the alloys have noncollinear

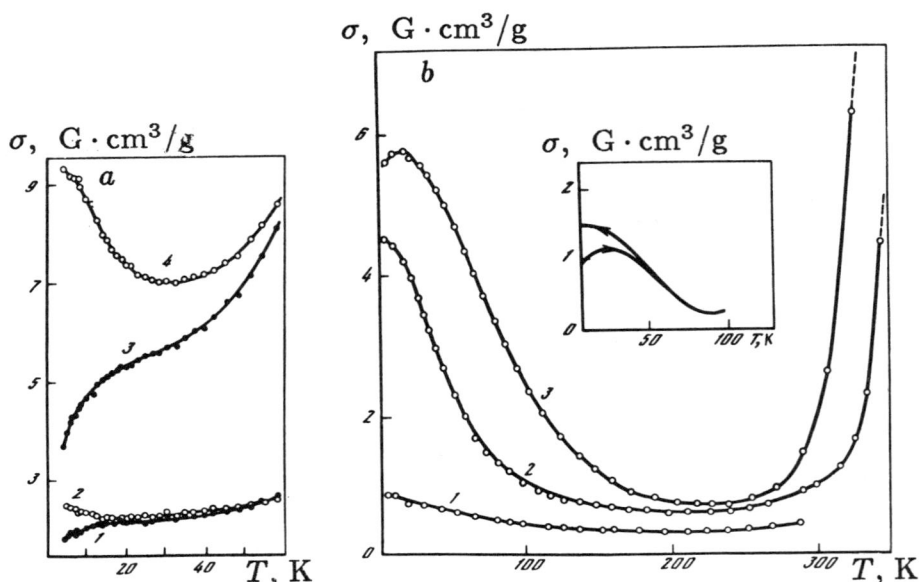

Figure 13　　Magnetization as a function of temperature for iron–platinum (a) and iron–rhodium (b) alloys of various composition measured after heating and cooling at constant magnetic fields (a) $H = 11\,kOe$ ((1, 2) $c_{Fe} = 32.2\,at\%$, (3, 4) $c_{Fe} = 32.2\,at\%$; (1, 3) cooling at $H = 0$, (2, 4) cooling at $H = 30\,kOe$); (b) $H = 240\,Oe$ (c_{Fe}, at%: (1) 50.5, (2) 50.75, (3) 51.0). Inset: Measurements at $H = 32\,Oe$ and $c_{Fe} = 50.75\,at\%$.

coplanar or noncoplanar magnetic structure with coexisting ferromagnetic and antiferromagnetic projections of the magnetic moments or whether the Fe–Rh alloys have a state like the "cluster glass embedded in an antiferromagnetic matrix," (where the clusters are, however, quite large and easily oriented by the magnetic field) as the Fe–Pt alloys do. The general results obtained favor the second hypothesis, as indicated by: 1) the presence of "residual" effects, especially temperature hysteresis upon cooling in a magnetic field (see Fig. 13b); 2) the existence of peaks in the susceptibility as a function of temperature characteristic of spin-glass freezing (Fig. 15) (the peaks for alloys with $c_{Fe} = 50.75$ and 51.0 at% are observed at temperatures of 30 and 40 K, respectively); external magnetic fields shift T_{max} to lower temperatures); and 3) the peak susceptibility χ_d is less than that for ferromagnetic samples of the same shape and size. For example, the maximum signal amplitude is approximately a factor of three smaller for the alloy with $c_{Fe} = 50.75$ at% than for the sample with 52 at% iron.

The presence of extreme magnetic inhomogeneity in Pt–Fe and Fe–Rh alloys during the transition from antiferromagnetism to ferromagnetism also produces kinetic effects. For example, the residual electrical resistance measured at 4.2 K is greatest for concentrations in this region (Table 4); a contribution proportional to $T^{3/2}$ (characteristic of spin glasses) appears in the electrical resistance as a function of temperature for the Fe–Rh alloys, while the Pt–Fe alloys are observed to have an anomalous electrical resistance as a function of temperature (kinetic effects in Pt–Fe alloys will be discussed in greater detail in the third paper in the present volume). A negative contribution to the magnetic resistance of Fe–Rh alloys appears in weak fields.

A study of an Fe–Rh monocrystal containing approximately 51.0 at% Fe indicated that there is practically no anisotropy in the magnetization at low temperatures. As in polycrystalline samples with an additional of iron, the monocrystal is observed to have a "weak ferromagnetic" phase at low temperatures. The ordering temperature for this phase was estimated from the $\sigma^2 = f(H/\sigma)$ construction:

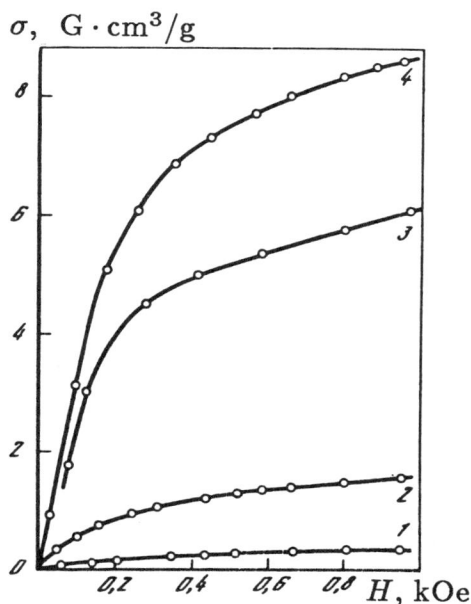

Figure 14   Magnetization of iron–rhodium alloys as a function of magnetic field strength in weak longitudinal fields at 4.2 K. c_{Fe}, at%: (1) 50.0; (2) 50.5; (3) 50.75; (4) 51.0.

$$\chi_d / \chi_{d,\max} \; (H = 0)$$

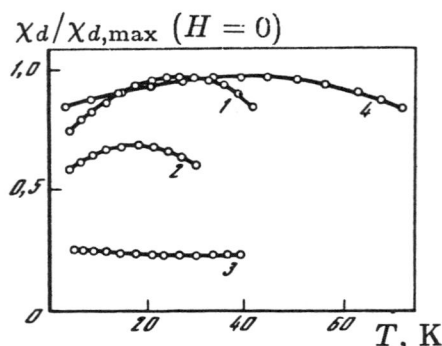

Figure 15   Relative differential magnetic susceptibility for iron–rhodium alloys as a function of temperature. (1–3) 50.75 at% Fe; (4) 51.0 at% Fe. H_l, Oe: (1) 0; (2) 25; (3) 100.

$T_{\text{ord}} \approx 40$ K. However, no maximum was observed in differential magnetic susceptibility as a function of temperature, and there was a plateau near 90 K.

Similar magnetic properties were observed in the ternary alloy $Fe_{49.5}Rh_{44}Ir_{6.5}$: increasing magnetization with decreasing temperature, the presence of a maximum in the differential magnetic susceptibility as a function of temperature, the differing behavior of the magnetization when the sample is cooled in- and outside of a magnetic field (Fig. 16). The Arrot construction implies that the ordering temperature for this alloy is 70 K. But there are also differences in the ternary alloy: the maximum in the susceptibility does not shift towards lower temperatures, but merely "flattens out" with increasing magnetic field strength.

It should be noted that the magnetization was observed to have a similar temperature dependence of the ternary alloy $FeRh_{0.9}Ir_{0.1}$ by Pál and coworkers [43], who found a spontaneous magnetic moment of $9 \, G \cdot cm^3 \cdot g^{-1}$ which was present up to 300 K. However, these values were not determined from the Arrot construction but by approximating the function $\sigma(H)$ as $H \to 0$, which yields higher values.

Pt–Fe Alloys with $c_{Fe} = 34$–36 at%   We shall now discuss the low-temperature properties of alloys with 34 and 36 at% Fe. As was shown in Section 1, even though the magnetization isotherms are of the characteristic form for ferromagnetic substances, the differential magnetic susceptibility as a function of temperature still indicates that there is an additional phase transition near 80 K. In order to determine the properties of these alloys, the magnetization was measured as a function of temperature in weak magnetic fields ($H = 160$ Oe) and in a 20-kOe field. The results of these measurements for the alloy with $c_{Fe} = 36$ at% are shown in Fig. 17 (the functions for the alloy with $c_{Fe} = 34$ at% are similar). The

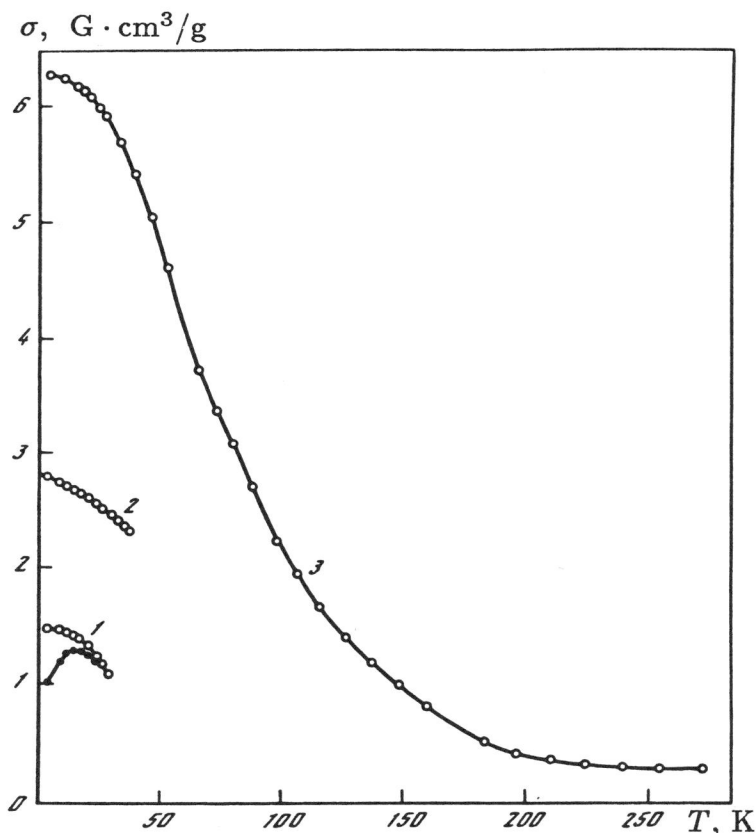

Figure 16   Magnetization as a function of temperature for the alloy $Fe_{49.5}Rh_{44}Ir_{6.5}$ for various fields H in Oersteds: (1) 9.6, (2) 32, (3) 240.

figure implies that the previous treatment of the samples has a substantial effect on the shape of the function $\sigma(T)$ at temperatures below 100 K: when the sample is cooled in a magnetic field, $\sigma(T)$ remains nearly constant down to $T = 4.2$ K; on the other hand, if the sample was cooled in a zero magnetic field, both the weak-field magnetization and the differential susceptibility decrease substantially with decreasing temperature.

The same behavior of the magnetization and susceptibility was observed in systems with binary magnetic transitions [44], and could be explained by the transition from the ferromagnetic state to a "cluster glass" state. When the samples

Table 4. Specific Electrical Resistance of Iron–Platinum and Iron–Rhodium Alloys at 4.2 K

c_{Fe}, at%	ρ, 10^{-6} Ohm $\cdot$ cm	c_{Fe}, at%	ρ, 10^{-6} Ohm $\cdot$ cm	c_{Fe}, at%	ρ, 10^{-6} Ohm $\cdot$ cm
	Pt–Fe	32.16	140	49.5	16
24.8	7.6	32.22	129	50.0	10
24.9	12	33.0	88	50.5	2.3
25.2	16.5		Fe–Rh	50.75	6.4
27.0	49	48.0	32	51.0	9.2
28.0	67	48.5	17	51.5	0.8
30.5	134	49.0	17	52.0	0.9
30.65	148				

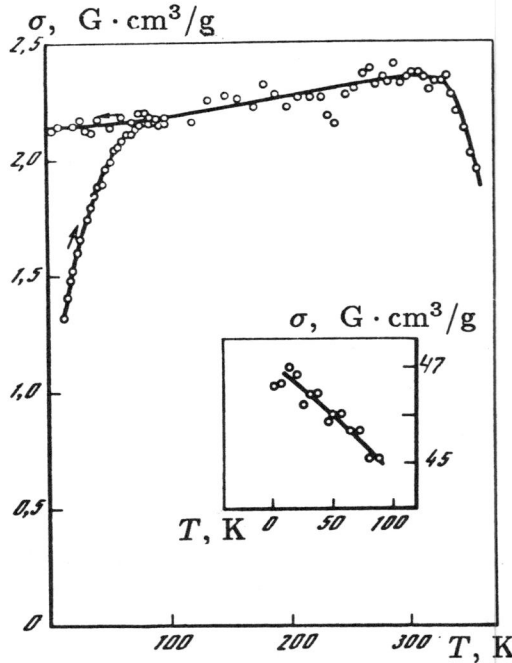

Figure 17 Temperature dependence of the magnetization for a polycrystalline iron–platinum alloy containing 36 at% Fe when heated and cooled in an 160 kOe magnetic field. Inset: magnetization as a function of temperature measured for heating in a 20 kOe field.

are cooled in a 4.2-kOe field, the hysteresis loops for samples with $c_{Fe} = 34\,at\%$ are shifted ($\Delta H_s \approx 50\,Oe$ for magnetization to $\pm 3.6\,kOe$); however, the hysteresis loop becomes symmetric once the sample has been magnetized to $\pm 5\,kOe$, i. e., the energy of unidirectional anisotropy is small in this alloy.

Since both antiferromagnetic and ferromagnetic reflections are present in the low-temperature ($T < 140\,K$) neutron diffraction patterns obtained by Bacon and Crangle [1] for the 36 at% Fe alloy, we can conclude that the low-temperature states of the alloys with $c_{Fe} = 34\text{–}36\,at\%$ need not be a simple "cluster glass" state. These states can apparently be characterized by a canted structure with possible "frozen components" along the third coordinate axis [12]. This is called a noncoplanar canted state. A similar state should also occur for Fe–Rh alloys on some interval of compositions with $51.0 < c_{Fe} < 51.5\,at\%$.

2.2. SOURCE FOR THE FORMATION OF CHAOTIC MAGNETIC STRUCTURES IN Pt–Fe AND Fe–Rh ALLOYS WITH COMPOSITIONS IN THE TRANSITION REGION FROM ANTIFERROMAGNETISM TO FERROMAGNETISM

The question of the existence of magnetic moments on the platinum and rhodium atoms in ordered Pt–Fe and Fe–Rh alloys is central to explaining the properties of and reasons for the formation of the various magnetic structures in these alloys.

We will first of all discuss the perfectly ordered stoichiometric-composition alloys Pt_3Fe and FeRh. At low temperatures, these alloys have antiferromagnetic ordering and magnetic structures described by wave vectors $\mathbf{k} = [1/2\ 1/2\ 0]$ for Pt_3Fe and $\mathbf{k} = [1/2\ 1/2\ 1/2]$ for FeRh (Figs. 18a and 18d), with the platinum and rhodium atoms being in a zero molecular field from the antiferromagnetically oriented iron atoms (in an ideal Pt_3Fe lattice, each platinum atom has two pairs of iron atoms with antiparallel moments among its nearest neighbors, while each rhodium atom in the FeRh lattice has four pairs of iron atoms with antiparallel magnetic moments as nearest neighbors); this is why neutron diffraction studies [1, 2, 6, 7] show that there are definitely no oriented magnetic moments on the rhodium and platinum atoms in ideal alloys of stoichiometric composition.

The low susceptibility and the absence of anomalies in the magnetization measured as a function of H at low temperatures in magnetic fields that are not too strong may be regarded as evidence of the fact that the platinum and rhodium atoms have absolutely no magnetic moments. At the same time, it is known from experiment that the platinum atoms in ordered and disordered Pt–Fe alloys have a magnetic moment of approximately $0.25\text{–}0.30\mu_B$ [2], while the rhodium atoms in ferromagnetic Fe–Rh alloys of equiatomic and nonequiatomic composition have a magnetic moment of $\sim 0.8\text{–}1.0\mu_B$ [5, 6]. The Pt atoms are observed to have a magnetic moment in several other alloys, such as, for example, Pt_3Cr [45], Pt_3Mn [46], and others. It is also well known that the matrix atoms in diluted

Fe, Co, and Ni alloys with a Pd, Pt, or Rh matrix become strongly polarized for even small concentrations of $3d$-metal impurities (0.1–0.3 at%) [47].

It is natural to assume that the platinum and rhodium atoms in concentrated Pt–Fe and Fe–Rh alloys might acquire magnetic moments under certain conditions. The primary reason for this would be the replacement of rhodium or platinum atoms by "additional" iron atoms in excess of the stoichiometric compositions Pt_3Fe and FeRh. Since the compensating molecular field condition is violated for the nearest-neighbor Pt and Rh atoms, the latter may acquire magnetic moments. We may assume, in accordance with the Jaccarino–Walker model [48], that the Pt and Rh atoms acquire magnetic moments when the number of iron atoms among their nearest neighbors reaches some critical value. A similar situation has been observed in the acquisition of magnetic moments by the Pd atoms in alloys of the form $(Pt_{1-x}Pd_x)_3Fe$ [49].

Although a theoretical description of the conditions for the formation of local magnetic moments of a certain size requires both an analysis of the variation of the electronic interaction parameters at a given lattice node and in its immediate vicinity and an analysis of the variation in the local density of states as a function of alloy composition [50], the present model under discussion should be a true reflection of the essence of the phenomena which occur.

In addition to polarizing the surrounding Pt or Rh atoms, the additional iron atoms may also lead to the establishment of ferromagnetic or canted structure over at least an elementary cell. A sort of complex or cluster with a nonzero total magnetic moment is formed in each case.

When there are few additional atoms, only isolated individual clusters will exist in the alloys. As the iron concentration increases, the clusters begin to overlap and "grow," which eventually leads to the formation of an "infinite cluster," i. e., chaotic ferromagnetic structure.

When magnetized at fairly high temperatures, the clusters will determine the superparamagnetic behavior of the alloy, since they are in a zero molecular field with respect to the main antiferromagnetic iron sublattice. With decreasing temperature, the Ruderman–Kittel–Kasuya–Yosida (RKKY) interaction will "freeze" the magnetic moments in random directions, thus forming a "cluster glass embedded in an antiferromagnetic matrix"-type structure. In the case where relatively large clusters are formed, both the exchange interaction and the magnetocrystalline anisotropy of the clusters will play an important role in the "freezing process" along with the RKKY interaction.

Cluster formation is a necessary consequence of the fact that the magnetic state of the alloy components depends on the chemical composition of the local environment [51]. The possible existence of clusters with giant magnetic moments in iron–rhodium alloys containing more than 50 at% iron has been confirmed through theoretical calculations [52].

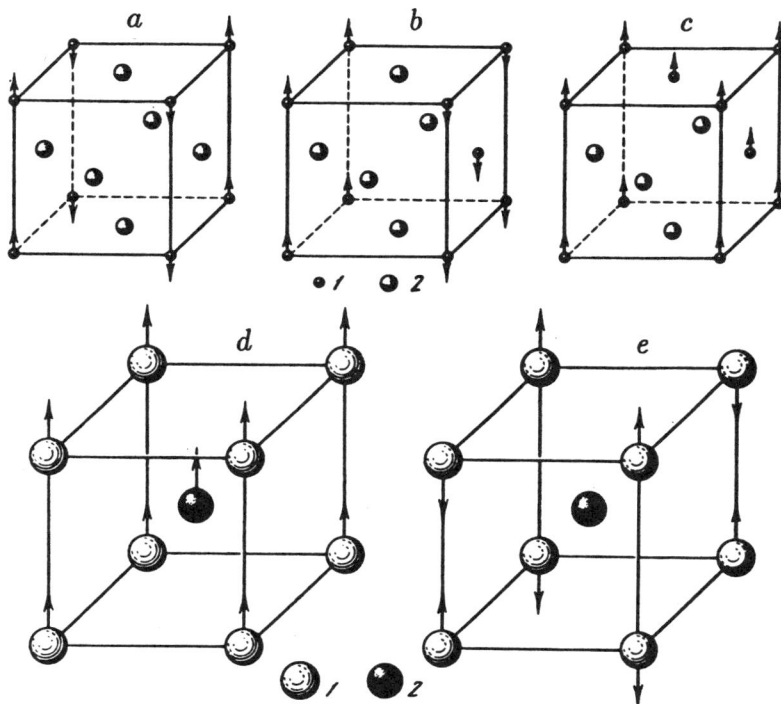

Figure 18 Magnetic structures of ordered iron–platinum and iron–rhodium alloys. (a)–(c) Pt–Fe alloys [2] ((a) antiferromagnetic structure with wave vector $\mathbf{k} = [1/2\ 1/2\ 0]$ in the stoichiometric-composition alloy Pt_3Fe; (b) antiferromagnetic structure with $\mathbf{k} = [1/2\ 0\ 0]$ in the $c_{Fe} \approx 31$ at% alloy; (c) ferromagnetic structure in the $c_{Fe} \approx 36$ at% alloy; (d), (e) equiatomic Fe–Rh alloy in the ferromagnetic (d) and antiferromagnetic (e) states [7] ((1) Fe; (2) Rh).

Estimates of the magnetic moments due to the additional iron atoms provide experimental confirmation for the formation of clusters. Measurements of the paramagnetic susceptibility imply that μ_{eff}/additional Fe atom is over $10\mu_B$ for the Pt–Fe alloys (see Section 1.3). The formation of clusters at $c_{Fe} = 50.75$ and 51.0 at% in Fe–Rh alloys is clearly implied by the values of the magnetic moment per "additional" iron atom presented below, which were determined from the magnetization in strong fields:

We shall now point out several differences due to crystal lattice topology between the cluster formation mechanisms in iron–rhodium and iron–platinum al-

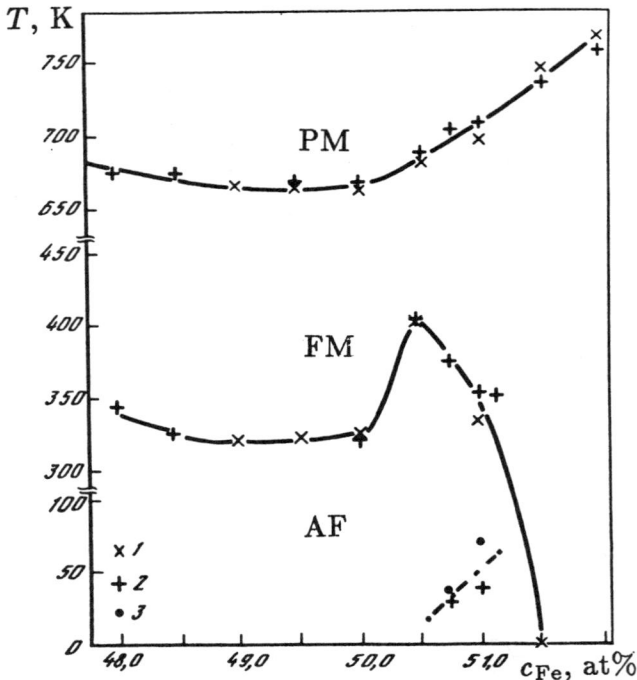

Figure 19   Magnetic $T-c$ phase diagram for iron–rhodium alloys. Phase transition temperatures determined from measurements of: (1) magnetization; (2) susceptibility; (3) residual magnetization; (PM) paramagnetic, (FM) ferromagnetic, and (AF) antiferromagnetic states.

loys. For example, in an ordered Fe–Pt alloy (with Cu_3Au-type ordering in the crystal structure), the additional iron atoms in excess of the stoichiometric composition, which replace the platinum atoms, sit on the faces of the cube, and have only the four iron atoms located at the vertices of these faces as nearest neighbors, and the ferromagnetic interaction between them leads to the formation of ferromagnetic sheets, i. e., to a change in the type of antiferromagnetic structure, while the conditions for the formation of giant clusters and the transition to ferromagnetism only exist when the deviation from stoichiometric composition is sufficiently large $(c_{Fe} \geq 30\,at\%)$ that the probability for the presence of two iron atoms on adjacent faces of an elementary cell is high enough. In iron–rhodium alloys, on the other hand, the additional iron atoms replacing the rhodium atoms at the center of the cube interact with eight iron atoms, which leads to "bulk ferromagnetism." In our view, this is what explains the difference in the widths of the transition region of compositions between antiferromagnetism and ferromagnetism for these alloys. For example, iron–platinum alloys become ferromagnetic at $c_{Fe} \approx 34\,at\%$, i. e., when there is one additional iron atom for every three to four cells. In iron–rhodium alloys, a deviation of as little as 1.5% from stoichiometric composition (one additional iron atom for every 33 cells) is sufficient for the establishment of ferromagnetic order in the alloy over the entire temperature range that magnetic order exists.

In conclusion, we present the magnetic T–c phase diagram obtained from our measurements for ordered Fe–Rh alloys of nearly equiatomic composition (Fig. 19). As the figure implies, the temperature for the transition from the ferromagnetic state to the paramagnetic state (a second-order phase transition) increases monotonically with increasing iron concentration in the alloys; the line for the phase transitions from the antiferromagnetic to ferromagnetic states is a non-monotonic function of alloy composition (the next paper in this volume will treat the properties of the phase transitions in iron–rhodium alloys). The transitions into the low-temperature chaotic magnetic structure that were observed are indicated in the phase diagram by a dash-dot line.

Thus, we have established the presence of small-scale magnetic ordering, including a completely new type of chaotic magnetic structure—"cluster glass embedded in an antiferromagnetic matrix"—in ordered iron–platinum and iron–rhodium alloys with compositions in the transition region from antiferromagnetism to ferromagnetism.

The existence of small-scale magnetic order is due to the induced internal exchange fields of the magnetic moments on the platinum and rhodium atoms when there additional iron atoms above the stoichiometric compositions Pt_3Fe and FeRh.

The authors express their appreciation to D. P. Rodionov for preparing the monocrystalline Pt–Fe alloy samples and to L. P. Maksimov, who kept the "Solenoid" facility in working order.

REFERENCES

1. Bacon, G. E., and J. Crangle, "Chemical and magnetic order in platinum rich Pt–Fe alloys," Proc. Roy. Soc. London A, vol. 272, pp. 387–405, 1963.
2. Men'shikov, A. Z., Yu. A. Dorofeev, V. A. Kazantsev, et al., "Magnetic structure of ordered iron–platinum alloys," Fiz. Metallov i Metallovedenie, vol. 38, pp. 505–518, 1974.
3. Vinokurova, L. I., I. N. Nikolaev, E. V. Mel'nikov, et al., "A study of ferro- and antiferromagnetic transformations in iron–platinum alloys using the Mössbauer effect," Fiz. Metallov i Metallovedenie, vol. 28, pp. 1098–1102, 1969.

4. Palaith, D., G. W. Kimball, R. S. Preston, et al., "Magnetic behavior of the PtFe system near Pt_3Fe," Phys. Rev., vol. 178, pp. 795–799, 1969.

5. Bertaut, E. F., A. Delapalme, F. Forrat, et al., "Magnetic structure work of the nuclear center of Grenoble," J. Appl. Phys., vol. 33 (supplement), pp. 1123–1124, 1962.

6. Shirane, G., C. W. Chen, P. A. Flinn, and R. Nathans, "Hyperfine fields and magnetic moments in the FeRh system," J. Appl. Phys., vol. 34, pp. 1044–1045, 1963.

7. De Bergevin, F., and L. Muldaver, "Antiferromagnetic–ferromagnetic transition in FeRh," J. Chem. Phys., vol. 35, pp. 1904–1905, 1961.

8. Vinokurova, L. I., V. Yu. Ivanov, and L. I. Sagoyan, "The transition from antiferromagnetism to ferromagnetism in ordered Pt–Fe alloys," Fiz. Metallov i Metallovedenie, vol. 49, pp. 537–552, 1980.

9. Izyumov, Yu. A., V. E. Naish, and R. P. Ozerov, Neitrony i Tverdoe Telo. T. 2. Neitronografiya magnetikov [Neutrons and Solids. Vol. 2. Neutron Diffraction Studies of Magnetic Substances], Atomizdat, Moscow, 1981.

10. Men'shikov, A. Z., Yu. A. Dorofeev, G. P. Gasnikova, et al., "Magnetic structure of ordered $Fe(Pd_{1-x}Au_x)_3$," Fiz. Metallov i Metallovedenie, vol. 47, pp. 1185–1189, 1979.

11. Dubinin, S. F., A. P. Vokhmyanin, and S. K. Sidorov, "Magnetic structure of the $Fe(Pt_{0.46}Pd_{0.52}Fe_{0.02})_3$ alloy with chaotic exchange interactions of various sign," Fiz. Metallov i Metallovedenie, vol. 51, pp. 966–974, 1981.

12. Medvedev, M. V., and A. V. Zaborov, "Magnetic states in a hardened Heisenberg magnet with chaotic nodes and competing exchange interactions. I. High-temperature magnetic states in simple and body-centered cubical lattices," Fiz. Metallov i Metallovedenie, vol. 52, pp. 272–284; "II. Magnetic states with partial spin-glass order in simple and body-centered cubical lattices," Fiz. Metallov i Metallovedenie, vol. 52, pp. 472–483, 1981.

13. Medvedev, M. V., and A. V. Zaborov, "Magnetic states in a hardened Heisenberg magnetic substance with chaotic nodes and competing exchange interactions. Phase diagrams for a magnetic substance with a fcc lattice," Fiz. Metallov i Metallovedenie, vol. 52, pp. 942–950, 1981.

14. Dubinin, S. F., S. G. Teploukhov, S. K. Sidorov, et al., "Magnetic structure of the Fe–Ni invar alloys," Phys. Status Solidi A, vol. 61, pp. 159–167, 1980.

15. Taplyugov, S. G., S. K. Sidorov, S. F. Dubinin, et al., "On antiferromagnetism of disordered Ni–Mn alloys," Phys. Status Solidi, vol. 21, p. K31–K34, 1974.

16. Kouvel, J. S., and J. B. Forsyth, "Ferromagnetic to canted-ferromagnetic transition in $Fe(Pd, Pt)_3$," J. Appl. Phys., vol. 40, pp. 1359–1361, 1969.

17. Vokhmyanin, A. P., V. V. Kelarev, A. N. Pirogov, et al., "Magnetic transformations in ordered $Fe_\alpha Mn_{1-\alpha}Pt_3$ alloys under changes in temperature," Fiz. Metallov i Metallovedenie, vol. 50, pp. 1010–1014, 1980.

18. Kadomatsu, H., B. Inoue, H. Fujii, et al., "Neutron diffraction study of ordered $Fe(Pd_{0.53}Pt_{0.47})$ single crystal," J. Phys. Soc. Japan, vol. 35, p. 1554, 1973.

19. Kohgi, M., Y. Ishikawa, and P. Radhakrishna, "Neutron scattering study of spin waves in an antiferromagnet $FePt_3$," Solid State Comm., vol. 27, pp. 409–411, 1978.

20. Crangle, J., "Some magnetic properties of platinum-rich Pt–Fe alloys," J. Phys. Radium., vol. 20, pp. 435–437, 1959.

21. Kohgi, M., and Y. Ishikawa, "Paramagnetic scattering of neutrons from a metallic antiferromagnet $FePt_3$," J. Phys. Soc. Japan, vol. 49, pp. 994–999, 1980.

22. Polovov, V. M., B. K. Ponomarev, and V. E. Antonov, "Some thermodynamical characteristics of the antiferro-ferromagnetism in iron–rhodium alloys," Fiz. Metallov i Metallovedenie, vol. 39, 977–985, 1975.

23. Bacon, G. E., and S. A. Wilson, "The unit cell dimensions of Pt_3Fe alloys," Proc. Phys. Soc., vol. 82, pp. 620–623, 1963.

24. Beaumont, R. H., H. Chihara, and J. A. Morrison, "An anomaly in the heat capacity of chromium at 38.5 °C," Phil. Mag., vol. 5, pp. 188–191, 1960.

25. Castaing, J., J. Danan, and M. Rieux, "Calorimetric and resistive investigation of the magnetic properties of CrB_2," Solid State Comm., vol. 10, pp. 563–565, 1972.

26. Tange, H., H. Tokunaga, and M.Goto, "Magnetic moment and Curie temperature of Ni–Mn disordered alloys," J. Phys. Soc. Japan, vol. 45, pp. 105–115, 1978.

27. Dorofeev, Yu. A., A. Z. Men'shikov, and G. A. Takzei, "Magnetic phase diagram for Fe_xCr_{1-x} alloys," Fiz. Metallov i Metallovedenie, vol. 55, pp. 948–954, 1983.

28. Burke, S. K., and B. D. Rainford, "The evolution of magnetic order in CrFe alloys," J. Phys. F: Metal Phys., vol. 13, pp. 441–482, 1983.

29. Sidorov, S. K., and S. F. Dubinin, "Atomic structure and ferromagnetism in ordered platinum–manganese alloys," Fiz. Metallov i Metallovedenie, vol. 24, pp. 859–867, 1967.

30. Men'shikov, A. Z., V. A. Kazantsev, and N. N. Kuz'min, "Magnetic state of iron–nickel–manganese alloys," Zh. Éksp. Teor. Fiz., vol. 71, pp. 648–656, 1976.

31. Men'shikov, A. Z., S. K. Sidorov, and A. E. Teplykh, "Magnetic state of γ-FeNiCr alloys with compositions in the critical region," Fiz. Metallov i Metallovedenie, vol. 45, pp. 949–957, 1978.

32. Tsiovkin, Yu. N., N. I. Kourov, and N. V. Volkenshtein, "Magnetic state of $(Pd_xPt_{1-x})_3Fe$," Fiz. Tverd. Tela, vol. 23, pp. 2614–2619, 1981.

33. Gasnikova, G. P., Yu. A. Dorofeev, A. Z. Mel'nikov, et al., "Magnetic phase transitions in ordered $(Fe_{1-x}Mn_x)Pt$ alloys, Fiz. Metallov i Metallovedenie, vol. 55, pp. 1138–1142, 1983.

34. Sadykov, R. A., P. L. Gruzin, V. N. Zaritskii, et al., "A neutron diffraction study of magnetic structure in substituted chalcospinels," Trudy Fiz. Inst. Akad. Nauk (Lebedev Phys. Inst.), vol. 139, pp. 150–168, 1982.

35. Adachi, K., K. Sato, and M. Takeda, "Magnetic properties of cobalt and nickel dihalcogenide compounds with pyrite structure," J. Phys. Soc. Japan, vol. 26, pp. 631–638, 1969.

36. Medvedev, M. V., "Two types of disordered ferromagnetic states in a Heisenberg magnet with random exchange bonds," Fiz. Tverd. Tela, vol. 21, pp. 3356–3364, 1979.

37. Medvedev, M. V., and S. M. Goryanova, "Asperomagnetism and short-range magnetic order in a Heisenberg magnet with random exchange bonds of different signs," Phys. Status Solidi B, vol. 97, pp. 415–419, 1980.

38. Kohgi, M., and Y. Ishikawa, "Magnetic excitations in a metallic antiferromagnet $FePt_3$," J. Phys. Soc. Japan, vol. 49, pp. 985–993, 1980.

39. Medvedev, M. V. and S. M. Goryanova, "Asperomagnetism and short-range magnetic order in a Heisenberg magnet with random exchange bonds of different signs," Phys. Status Solidi B, vol. 98, p. 143, 1980.

40. Tholence, J. L., and R. Tournier, "Remanent magnetism of spin glasses," Physica B, vol. 86/88, pp. 873–874, 1977.

41. Mydosh, J. A., "Spin glasses and mictomagnets," Proc. AIP Conf., vol. 24, pp. 131–137, 1975.

42. Beck, P. A., "Properties of mictomagnets (spin glasses)," Prog. Mat. Sci., vol. 23, pp. 1–49, 1978.

43. Pál, L., G. Zimmer, M. P.-Horváth, and J. Paitz, "Impurities in antiferromagnets at lattice sites of zero exchange field," J. Phys., vol. 32 (supplement), pp. C1-861–C1-862, 1971.

44. Nieuwenhuys, G. J., B. H. Verbeek, and J. A. Mydosh, "Towards a uniform magnetic phase diagram for magnetic alloys with mixed types of order," J. Appl. Phys., vol. 50, pp. 1685–1690, 1979.

45. Burke, S. K., B. D. Rainford, D. E. G. Williams, et al., "Magnetization density in ferrimagnetic Pt_3Cr," J. Magnetism Magnetic Materials, vol. 15/18, pp. 505–506, 1980.

46. Antonini, B., F. Lucari, F. Menzinger, et al., "Magnetization distribution in ferromagnetic $MnPt_3$ by a polarized-neutron investigation," Phys. Rev., vol.187, pp. 611–618, 1969.

47. Clogston, A. M., B. T. Matthias, and M. Peter, et al., "Local magnetic moment associated with an iron atom dissolved in metal alloys," Phys. Rev., vol. 125, pp. 541–552, 1962.

48. Jaccarino, V., and L. R. Walker, "Discontinuous occurrence of localized moments in metals," Phys. Rev. Lett., vol. 15, pp. 258–261, 1965.

49. Savchenkova, S. F., Osobennosti Magnitnykh i Élektricheskikh svoistv sistem $Fe(Pd_xPt_{1-x})_3$ [The magnetic and electrical properties of alloys in the system

Fe$(Pd_xPt_{1-x})_3$], Candidate's Dissertation in Phys.-Math. Sci., Institute for Metal Physics, Ural Science Center, USSR Academy of Sciences, Sverdlovsk, 1972.

50. Zaborov, A. V., and M. V. Medvedev, "A model for a dilute ferromagnet with magnetic moments depending on the local environment," Fiz. Metallov i Metallovedenie, vol. 53, pp. 651–660, 1982.

51. Zaborov, A. V. and M. V. Medvedev, "A binary ferromagnetic alloy with magnetic moments dependent on local environment," Phys. Status Solidi B, vol. 116, pp. 511–523, 1983.

52. Khan, M. K., J. Khwaja, and C. Demengeat, "Origin of giant moments in non-stoichiometric FeRh alloys," J. Phys, vol. 42, pp. 573–577, 1981.

INSTABILITY OF MAGNETIC STATES IN ORDERED IRON–PLATINUM AND IRON–RHODIUM ALLOYS

V. G. Veselago, L. I. Vinokurova, A. V.
Vlasov, N. I. Kulikov, B. K. Ponomarev,
M. Pardavi-Horváth, and L. I. Sagoyan

Abstract Instabilities were observed in the magnetic states which form in iron–rhodium and iron–platinum alloys at strong magnetic fields and high pressures. It was shown that the observed instabilities are due to certain characteristics of the electron band structure of the alloys.

Introduction

Research on the physical properties of ordered iron–platinum and iron–rhodium alloys has shown that the magnetic states occurring in these alloys turn out to be quite sensitive to both variations in alloy composition and the influence of external thermodynamic parameters (temperature, magnetic field, pressure).

The first paper in this volume was devoted to a determination of the T–c magnetic phase diagram for Pt–Fe alloys with Fe concentrations between 24.8 and 36 at% (see Fig. 3 in the previous paper). Stoichiometric-composition alloys of the form Pt_3Fe were found to be antiferromagnetic. A first-order phase transition from the antiferromagnetic state to the ferromagnetic state is observed in Fe–Rh alloys at $T_{cr} \approx 330\,K$ as the temperature increases. The temperature of the phase transition T_{cr} depends on the composition of the alloy, and an analougous sitiation holds for Pt–Fe alloys with compositions in the transition region between antiferromagnetism (AFM) and ferromagnetism (FM): the alloys become ferromagnetic with increasing temperature at $T_1 \approx 100\,K$ within a narrow range of iron concentrations ($30.65 \leq c_{Fe} \leq 32.3$ at%).

The two alloy systems are observed to have common traits: strong magnetic fields stabilize the ferromagnetic state, while pressure stabilizes the antiferromag-

45

netic state. The present paper will present a detailed discussion of the nature
of and reasons for the instability of the magnetic states with respect to strong
magnetic fields and high pressures.

1. MAGNETIC PHASE TRANSITIONS INDUCED BY AN EXTERNAL MAGNETIC FIELD IN IRON–PLATINUM AND IRON–RHODIUM ALLOYS

The temperature dependence of H_{cr} (H_{cr} is the magnetic field strength at which
the first-order phase transition from the antiferromagnetic state into the ferromagnetic state takes place for a given temperature) has been studied numerous times
both in strong constant fields (as large as 80 kOe) [1] and in pulsed magnetic
fields [2–4]. There is, however, still some disagreement on the nature of this function. This is due to the fact that at low temperatures, the critical fields are quite
large, so that the constant-field measurements only yield information on the function $H_{cr}(T)$ over a small range of temperatures, while the magnetocaloric effect
must be taken into account in pulsed-field measurements.

In our experiments, it was established for the first time that the antiferromagnetic state in iron-platinum alloys of non-stoichiometric composition ($c_{Fe} \geq$
25.2 at%) is also unstable to strong magnetic fields, while the variation in H_{cr} for
the Pt–Fe alloys as a function of temperature turned out to be a function of alloy
composition [5].

1.1. MAGNETIC PROPERTIES OF IRON–PLATINUM ALLOYS IN STRONG MAGNETIC FIELDS

Magnetization measurements of monocrystalline Pt–Fe alloys in static magnetic
fields of up to 150 kOe indicate that for $T < T_1$, there is a first-order metamagnetic phase transition at $H = H_{cr}$. This phase transition leads to a sharp increase
in the magnetization, with the samples becoming either weakly ferromagnetic
($c_{Fe} < 30.5$ at%) or ferromagnetic ($c_{Fe} \geq 30.5$ at%). Figure 1 shows the magnetization as a function of magnetic field strength at various temperatures for
several of the crystals included in the study. In samples of almost-stoichiometric
composition ($c_{Fe} = 24.8 \pm 0.1$ at%), the magnetization is a nearly linear function
of the field, as is generally the case for antiferromagnets. There are no jump discontinuities in the $\sigma(H)$ magnetization curves for $4.2 < T < 200$ K ($T_N = 170$ K).
However, the addition of a small amount of iron in excess of the stoichiometric
composition Pt_3Fe makes the low-temperature phase ($T < T_1$) unstable. Once
the external magnetic field strength reaches the critical value, the alloy undergoes
a transition into a new phase with nearly the same magnetization as for $T > T_1$

(if the appropriate corrections are made for the temperature dependence of the magnetization (Fig. 1a)).

Note that the function $\sigma(H)$ is nonlinear above the discontinuity. Similar nonlinearity is also observed for $T > T_1$ in strong magnetic fields. However, the alloys with $c_{Fe} < 30.5$ at% still have high susceptibility, even in strong magnetic fields, and no approach to saturation is observed. On the other hand, the alloys with $c_{Fe} \geq 30.5$ at% have low susceptibility ($\sim 2.5 \cdot 10^{-4}$) in strong magnetic fields, as is generally the case for antiferromagnets; however, the spontaneous magnetization $\sigma_0 = (0.5-0.6)\sigma_{0,disord}$, where $\sigma_{0,disord}$ is the spontaneous magnetization for a disordered alloy of the same composition.

The critical magnetic field H_{cr} is shown as a function of temperature in Fig. 2. The function H_{cr} is a power law, i. e., $H_{cr}^2 = a(T - T_1)$, for alloys with $c_{Fe} < 30.5$ at%. We see from Fig. 2a that the curves for the alloy with $c_{Fe} = 25.2$ at% have constant slope over the temperature interval studied ($77\,K < T < T_1$), while the slope changes at $T \approx 90\,K$ for the alloy with $c_{Fe} = 28$ at%.

The critical field values were determined from the functions $\sigma(H)$ at constant temperature. Note that these values are identical to those determined from the maximum slope of the curves $\sigma(T)_{H=const}$. Since H_{cr} increases rapidly with decreasing temperature, it turned out that for constant magnetic fields of up to 150 kOe, induced magnetic phase transitions in alloys with small amounts of excess iron can only be observed on a small temperature interval ($60\,K < T < T_1$).

H_{cr} shows different behavior as a function of temperature for alloys with $c_{Fe} \geq 30.5$ at%: H_{cr} starts to increase at roughly $T = 25\,K$. Significant hysteresis (which is especially large at low temperatures ($T < 50\,K$)) was also observed. For example, the amount of hysteresis, i. e., the difference between the values of H_{cr} when the field is on the increase and when it is on the decrease ($\Delta H = H_{cr}(\uparrow) - H_{cr}(\downarrow)$) is 35.8 and 43.6 kOe, respectively, at $T = 4.2\,K$ and $1.95\,K$ for the $c_{Fe} = 32.16$ at% alloy.

The variation in the critical field with temperature for samples with iron concentration greater than or equal to 30.5 at% can be understood if we look at the low-temperature magnetic structure of the alloys. The magnetization measurements imply that these alloys are very inhomogeneous at low temperatures and consist of an antiferromagnetic matrix with "frozen" spin-glass-like ferro- or ferrimagnetic cluster inclusions. This type of cluster order is customarily called a "cluster glass." Experiments indicate that that cooling in a constant magnetic field from temperatures above the "freezing point" T_f or the prior imposition of a strong magnetic field leads to partial ordering of the clusters (residual magnetization $\sigma_r \neq 0$), i. e., it is as though a magnetized "cluster glass" had been formed. In this case, the total magnetic field in the sample will consist of the internal field set up by the magnetized clusters plus the external field. This is what in fact causes the critical magnetic field to decrease with decreasing temperature.

The fact that the hysteresis increases with decreasing temperature (for $T <$

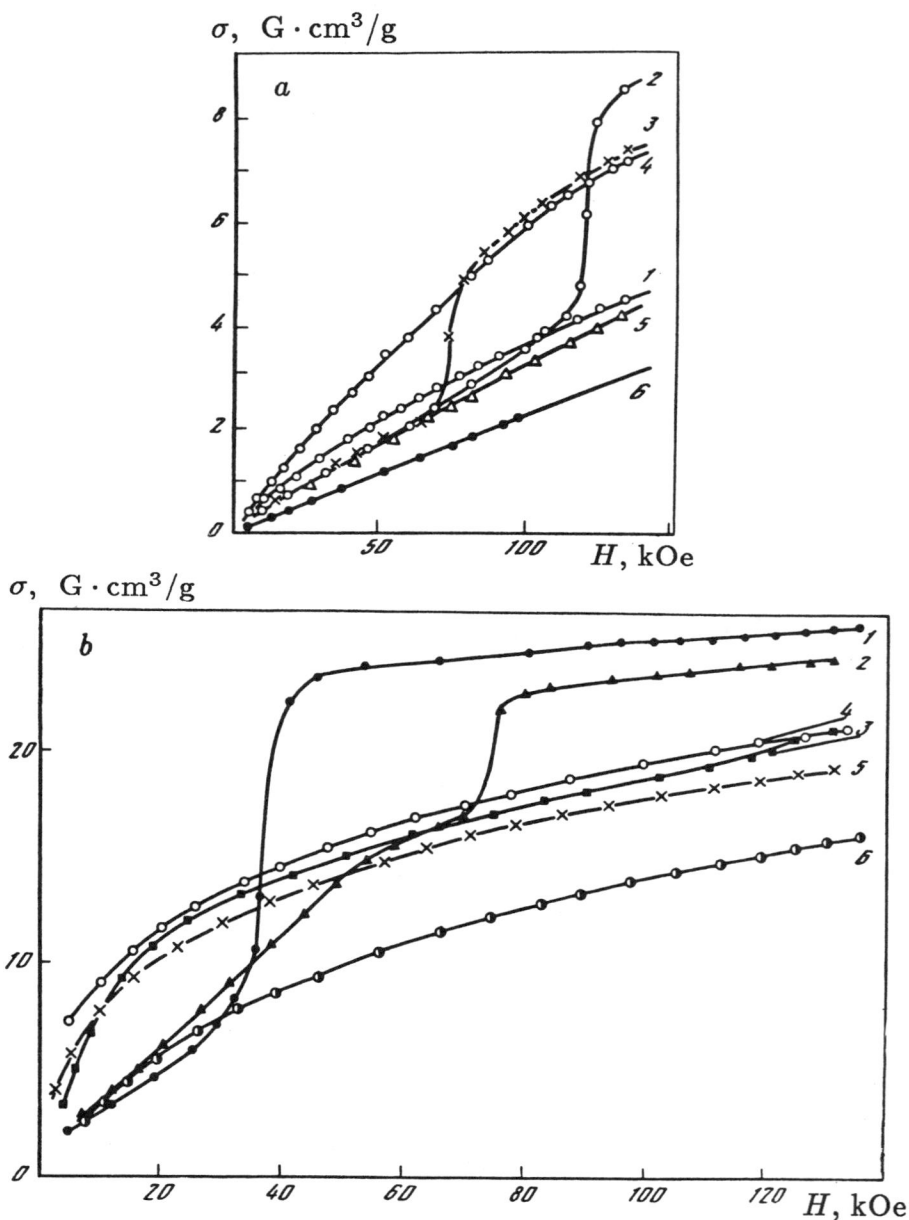

Figure 1   Specific magnetization as a function of magnetic field strength for iron–platinum alloys. (a) $c_{Fe} = 27\,at\%$ ((1) $T = 4.2$, (2) 77, (3) 95, (4) 106, (5) 154.5, and (6) 290 K); (b) $c_{Fe} = 30.5\,at\%$ ($c_{Fe} = 30.5\,at\%$ ((1) $T = 4.2$, (2) 77, (3) 108.5, (4) 125, (5) 154, and (6) 211.5 K).

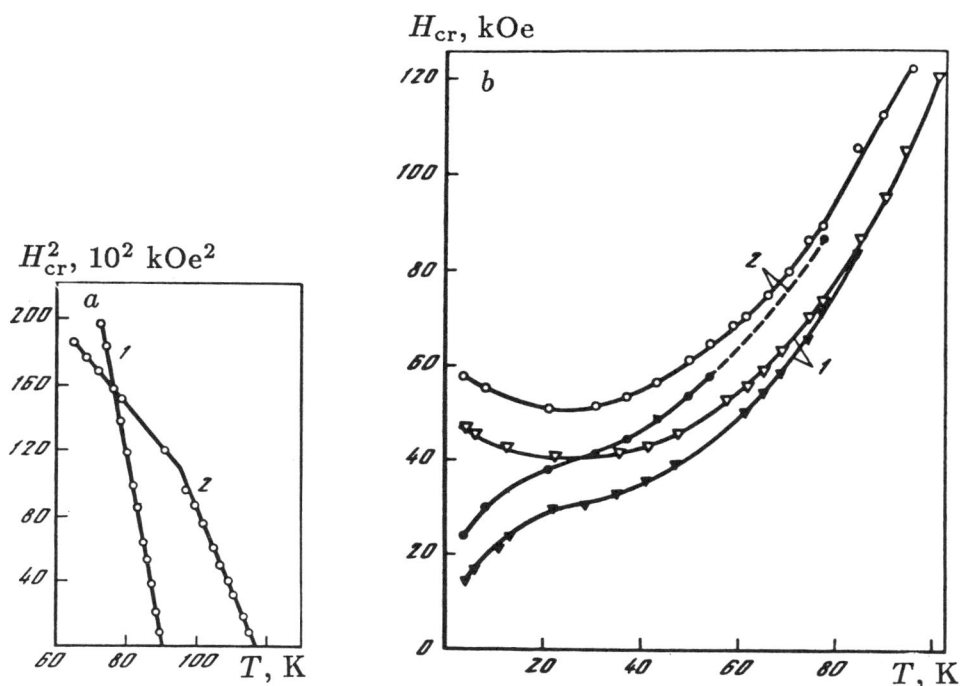

Figure 2 H_{cr} as a function of temperature for Pt–Fe alloys. (a): (1) 25.2 at% Fe, (2) 28 at% Fe; (b) 32.2 at% Fe ((1) $H\|[001]$, (2) $H\|[111]$—filled symbols correspond to decreasing magnetic field, and open symbols correspond to increasing magnetic field).

50 K) is also due to a similar relationship between critical field and magnetic structure: As the external field decreases, the internal field orients itself in the opposite direction.

The increased magnetic field at temperatures above T_f in these alloys is obviously also due to their magnetic structure. The higher the temperature, the more difficult it is for the external field to orient the clusters and the smaller the contribution from the internal magnetic field to the total field.

The temperature dependence of the size of the discontinuity in magnetization for $H = H_{cr}$ in the vicinity of T_1 can be written in the following form: $\Delta\sigma^2 = b(T_1 - T)$, where T_1 is the temperature of the phase transition at zero magnetic field.

The change in entropy during a first-order magnetic phase transition can be determined from the size of the discontinuity in magnetization and the temperature dependence of the critical field using the Clausius–Clapeyron equation:

$$(\Delta S)_H = -\Delta \sigma_S (dH_{cr}/dT)_p. \tag{1}$$

This information is given for $T = 77$ K in Table 1.

The quantity ΔS has a negative sign for alloys with high iron concentration because of the fact that H_{cr} increases with temperature at liquid-nitrogen temperatures. The quantity ΔS increases somewhat as the temperature increases and approaches T_1. Thus, for example, $\Delta S = 0.78 \, \text{mJ} \cdot \text{g}^{-1} \cdot \text{K}^{-1}$ for the alloy with $c_{Fe} = 28$ at% at $T = 105$ K, which is similar to the value obtained for ΔS from measurements of the heat capacity of this sample $(0.9 \, \text{mJ} \cdot \text{g}^{-1} \cdot \text{K}^{-1})$. It should be noted that the change in entropy during the magnetic-field-induced phase transitions is an order of magnitude smaller in Pt–Fe alloys than in Fe–Rh alloys.

Under adiabatic conditions, the size of the magnetocaloric effect can be calculated using the formula

$$(\Delta T)_s = -T(\Delta S)_T / C_H, \tag{2}$$

where C_H is the heat capacity at constant field.

Table 1 gives the values of ΔT at 77 K calculated for Pt–Fe alloys containing 28 and 32 at% iron, as well as the ΔT values experimentally determined for the phase transitions at $H = H_{cr}$. As the table indicates, the signs of the effect are in agreement with those expected from the thermodynamical equations, and the values are in reasonable agreement.

Since the constant magnetic field values turned out to be too low to establish whether induced magnetic phase transitions into a weakly ferromagnetic or ferro-

Table 1. Change in Entropy and Calculated and Experimental Values of the Magnetocaloric Effect ΔT for Iron–Platinum Alloys for a Phase Transition in Field $H = H_{cr}$ at 77 K

c_{Fe}, at%	ΔS, mJ $\cdot$ g$^{-1} \cdot$ K^{-1}	ΔT, K	
		Calculation	Experiment
25.2	0.57		
27.0	0.65		
28.0	0.37	-0.30	-(0.4 ± 0.1)
30.5	-0.71		
32.2	-0.62	0.49	0.4 ± 0.1

Table 2. Parameters Determined from Magnetization Measurements in Pulsed Magnetic Fields at 4.2 K

Parameter	c_{Fe}, at%				
	25.2	27	28	30.5	32.2
Direction of magnetization	[100]	[001]	[110]	[110]	[001]
$H_{cr}(\uparrow)$, kOe	232	195	157	82	42
$H_{cr}(\downarrow)$, kOe	218	148	132	48	13.5
σ'_0, G $\cdot$ cm$^3 \cdot$ g^{-1}	0.4	8.8	12.0	32.0	33.5
μ'_0/alloy atom, μ_B	0.01	0.25	0.34	0.9	0.9
μ'_0/Fe atom, μ_B	0.05	0.9	1.2	2.9	2.8
σ''_0, G $\cdot$ cm$^3 \cdot$ g^{-1}	0	6.8	10.4	32.3	33.4
μ''_0/alloy atom, μ_B	0	0.2	0.3	0.9	0.9
$\sigma_{300\,kOe}$, G $\cdot$ cm$^3 \cdot$ g^{-1}	10.0	12.6	15.5	34.4	35.0
$(\partial\sigma/\partial H)_{300\,kOe}$, 10^{-5} cm$^3 \cdot$ g^{-1}	3.2	1.2	1.2	0.6	0.4
$\Delta\sigma$, G $\cdot$ cm$^3 \cdot$ g^{-1}	2.8	5.0	7.2	15.0	7.0
μ_0/alloy atom, μ_B	0.08	0.14	0.20	0.40	0.20
μ_0/Fe atom, μ_B	0.32	0.52	0.72	1.35	0.60
μ_0/add'l Fe atom, μ_B	40.5	7.04	6.75	7.50	2.64

NOTE: σ_0 is the spontaneous magnetization and $\Delta\sigma$ the jump in magnetization at $H = H_{cr}$. The quantities σ'_0 and μ'_0 were determined by extrapolating $\sigma(H)$ for $H > H_{cr}$ to $H = 0$, while the quantities σ''_0 and μ''_0 were determined from the values of the function $\sigma^2 = f(H/\sigma)$ at $H > H_{cr}$, and μ was determined from $\Delta\sigma$.

magnetic phase existed in all of the alloys, the magnetizations of the monocrystals were measured at liquid-helium and liquid-nitrogen temperatures in pulsed (0.01-sec long pulses) magnetic fields of up to 350 kOe. The measurements were made inductively. The relative error in the magnetization measurements, $\Delta\sigma/\sigma$, was 5–8%. Absolute field values were determined to within 8% using an induction coil and an RC integrator. The measurements were carried out on the same samples used in the constant-field measurements.

Figure 3 shows the results of measurements on crystals with $c_{Fe} = 24.8\text{--}28.0\,\text{at}\%$ at 4.2 K. As may be seen from the figure, the function $\sigma(H)$ for the alloy with $c_{Fe} = 24.8\,\text{at}\%$ becomes nonlinear for very strong fields $(H \approx 200\,\text{kOe})$. It is not clear whether this is due to the fact that the field strength is approaching the critical value for the transition from antiferromagnetism to ferromagnetism, or due to the fact that induced moments have appeared on the platinum atoms. Extrapolation of $\sigma(H)$ to $H = 0$ yields the finite value $\sigma_0 = 0.2\,\text{G} \cdot \text{cm}^3/\text{g}$.

The experimentally measured values of H_{cr}, the size of the discontinuity in magnetization at H_{cr} $(\Delta\sigma)$, the spontaneous magnetization (σ_0) (both that determined by extrapolating $\sigma(H)$ to $H = 0$ and that determined by constructing the functions $\sigma^2 = f(H/\sigma)$), and the magnetic susceptibilities at 300 kOe $(\partial\sigma/\partial H)_{300\,\text{kOe}}$ are given in Table 2, which also contains the results of a calculation of the magnetic moments per "alloy atom" from the spontaneous magnetization. The change in magnetic moment per iron atom (and per "additional" iron atom) resulting from the phase transition were calculated from the discontinuity in magnetization at $H = H_{cr}$.

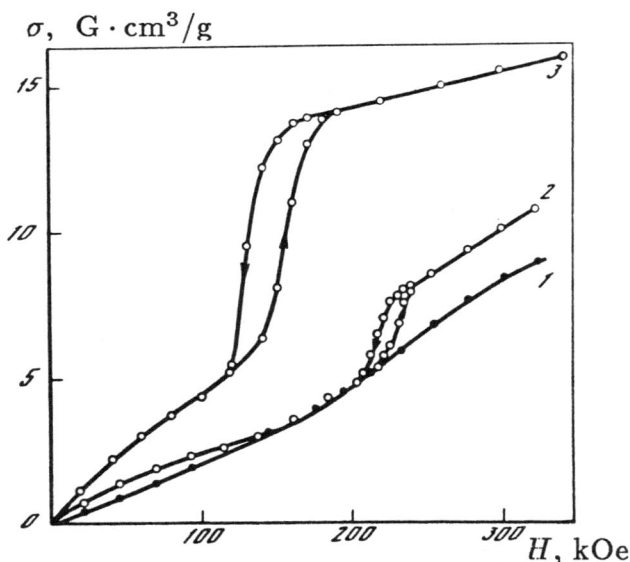

Figure 3   Specific magnetization as a function of magnetic field strength measured in pulsed magnetic fields at 4.2 K. c_{Fe}, at%: (1) 24.8, (2) 25.2, (3) 28.

1.2. ANISOTROPY OF THE CRITICAL FIELD.
THE MAGNETIC PROPERTIES OF Pt–Fe ALLOYS

In the process of studying the critical fields in iron–platinum alloy monocrystals, we noticed that they were anisotropic. In order to avoid the errors associated with possible differences in composition between samples cut from different regions of the monocrystal (as in the parallelepipedal samples used in the magnetization measurements), the measurements were carried out on cubical samples of side length 1.7–2.7 mm.

The measurements indicated that for small deviations from stoichiometric composition ($c_{Fe} = 25.2$ at%) the alloys were nearly isotropic; the magnetization and critical field strength show practically no variation with crystal orientation. A completely different pattern was observed in the samples with higher iron content: an increased Fe concentration leads to anisotropy in the critical field values; however, none of the alloys studied with Fe contents of 28 at% or less had any anisotropy in the magnetization at weak fields of order a few kilo-oersteds [5].

The anisotropy in the samples with 27–28 and 30–32 at% Fe is different in nature; in the former case, the maximum critical field occurred along axes of the type < 110 >; in the latter case, this occurred for axes of the type < 111 >. Experiments on samples with high iron concentration ($c_{Fe} \geq 30.5$ at%) showed no change in the behavior of the anisotropy as the temperature was decreased to that of liquid helium: H_{cr} at 4.2 K is largest along the < 111 > axis (Fig. 4).

Galvanomagnetic effects turned out to be more sensitive than the magnetic measurements. Measuring the "rotation curves" also enabled us to determine the

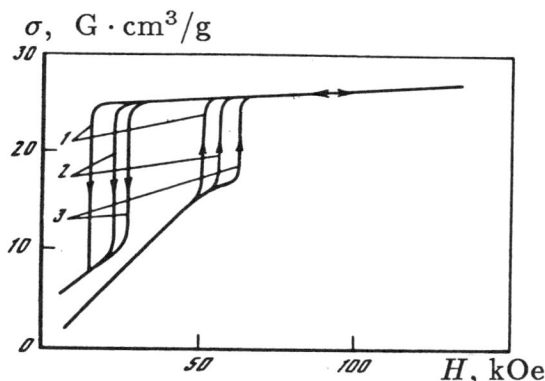

Figure 4 Specific magnetization of a monocrystalline Pt–Fe alloy with Fe content 32 at% at 4.2 K as a function of the magnetic field strength for magnetic fields applied along various crystallographic axes. (1) H∥[100]; (2) H∥[110]; (3) H∥[111].

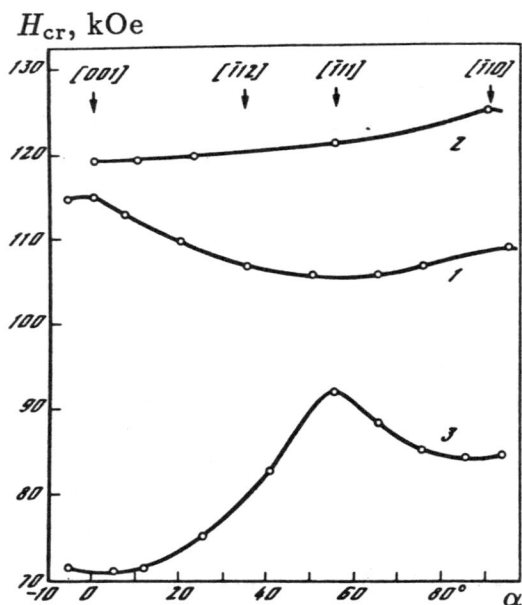

Figure 5 H_{cr} for iron–platinum alloy monocrystals at 77 K as a function of the angle between the magnetic field and the [011] crystallographic axis as the field is moved around the
(110) plane. c_{Fe}, at%: (1) 25.2; (2) 28.0; (3) 30.5.

anisotropy in H_{cr} for the samples with 25.2 at% Fe: we measured the electrical
resistance ρ in samples of order 7–7.5 mm in length with square cross sections (0.7×
0.7 mm). The values of H_{cr} were determined from the discontinuities in resistivity
associated with the sharp increase in sample magnetization. Figure 5 shows the
variation in the critical field at 77 K for monocrystals of various composition as
the magnetic field is rotated in the (110) crystallographic plane. In contrast to
the magnetization data, H_{cr} is also observed to be anisotropic at $c_{Fe} = 25.2$ at%,
with H_{cr} being largest along the $< 100 >$ axis. However, the anisotropy in H_{cr},
i. e., the difference between the critical fields along the [001] and [111] axes, is less
than 10%, as opposed to the values of 20–30% measured in alloys with higher iron
content. These results imply that neither decreasing the temperature to 4.2 K nor
exposure to strong magnetic fields affects the behavior of the anisotropy.

Thus, it was found that the direction in which the critical field is maximum
varies with increasing iron concentration: it appears to rotate in an {110}-type
plane from the [001] crystallographic axis for $c_{Fe} = 25.2$ at% to the [110] axis at
$c_{Fe} = 28$ at%, finally coming back to an intermediate position along the [111] axis.

It may be assumed that the observed variations in the properties of the anisotropy as a function of iron concentration are due to the fact that the crystals have anisotropy constants whose absolute value and (probably) sign depend on composition. The anisotropy in magnetically-ordered metals and alloys is generally thought of as being due to the spin-orbit interaction [6]. This interaction means that the presence of spontaneous magnetization (or changes in the spontaneous magnetization) causes changes in the band structure. The transitions leading to the nonzero spontaneous magnetic moment observed in strong magnetic fields are also accompanied by changes in the band structure of the crystals, and, in particular, changes in the populations of the spin subbands.

1.3. PROPERTIES OF PHASE TRANSITIONS IN Pt–Fe ALLOYS AT STRONG MAGNETIC FIELDS

In spite of the fact that many properties of transition metals and alloys can be successfully explained in the localized magnetic moment model approximation, it is nevertheless well known that the $3d$ electrons in these materials may become more weakly localized, especially in near the critical concentrations [7].

The $3d$ elements show band characteristics even in very diffuse alloys, because of the strong overlap between their electron wave functions and the d wave functions of the matrix, as in Rh–Co and Ru–Fe alloys, for example. At sufficiently high concentrations, a transition to large-scale ferromagnetic or antiferromagnetic order occurs; this order can only be described using the band model (a collectivized-electron model). It is natural to assume that some of the observed properties of the Pt–Fe system might find explanations within the framework of this model.

Indeed, Sumiyama and Graham [8] found that the coefficient of thermal expansion α in the stoichiometric-composition alloy was negative at low temperatures (less than or equal to 50 K). Analysis of the various contributions to the coefficient of thermal expansion implied that the "magnetic contribution" to α is likewise negative and proportional to temperature for low temperatures, in agreement with Wohlfarth's theory of collective magnetism [9].

Our experiments also indicated that the coefficient of thermal expansion was negative at low temperatures for alloys having compositions in the transition region.

It is well known that an analysis of the magnetization as a function of temperature and magnetic field strength using the Edwards–Wohlfarth equation [10]

$$\sigma^2(H, T) = 2\chi_0\sigma_0^2\frac{H}{\sigma(H, T)} + 1\left[1 - \left(\frac{T}{T_C}\right)\right]\sigma_0 \qquad (3)$$

can be used as a criterion for determining whether band electron models can be used for a given transition metal or alloy.

This equation implies that the functions $\sigma^2 = f(H/\sigma)$ should consist of parallel straight lines (deviations from straight lines can only occur in magnetically inhomogeneous materials). The lines of equal magnetization $\sigma(H, T) = \text{const}$ should be parabolas in the (T, H) plane or straight lines in the (T^2, H) plane.

Besnuns, et al. [11] made improvements to the Wohlfarth–Stoner theory and took additional, higher-order terms into account in the Stoner equation [12], which led to the following equation for the magnetization:

$$H/\sigma = a_{10} + a_{12}T^2 + a_{14}T^4 + a_{32}\sigma^2 T^2, \tag{4}$$

where the a_{ij} are coefficients determined by the interaction between the collectivized electrons, the density of states on the paramagnetic Fermi surface, and the derivatives of the density of states with respect to energy. This equation (like the Edwards–Wohlfarth equation) also implies a linear relation for $\sigma^2 = f(H/\sigma)$. However, taking the additional terms into account revealed additional properties of the band magnetism criteria. In particular, it turned out that the Landau A and B coefficients are no longer constants (B is the slope of the lines in the equation $\sigma^2 = f(H/\sigma)$ and A is their intercept on the H/σ axis), but depend quadratically on temperature over the region where the band magnetism criterion is applicable.

Our magnetization data imply that the lines of equal magnetization are nearly straight lines in the (H, T^2) plane for alloys with high iron content, which indicates that the idea of collectivized magnetic moment carriers can be used, at least for these alloys. In addition, we found that the experimental functions $\sigma^2 = f(H/\sigma)$ for all of the alloys are straight lines at strong magnetic fields.

This data leads us to believe that the band structure of the alloy and, consequently, the density of states in the Fermi surface, changes as the Pt_3Fe is diluted by iron. The variation in the coefficient in front of the linear (in temperature) term in the heat capacity (which determines the electron heat capacity and is associated with the density of states in the Fermi level) is an indication of the variation in the density of states. This coefficient is much larger for the alloy with $c_{Fe} = 32\,\text{at\%}$ than for the alloy with $c_{Fe} = 28\,\text{at\%}$ (9.0 and 2.8 mJ $\cdot$ mole^{-1} $\cdot$ K^{-1}, respectively). However, magnetic inhomogeneities may also contribute substantially to this coefficient at compositions in the vicinity of the transition to ferromagnetism [13].

These variations may apparently lead to the onset of weak ferromagnetism in the collectivized system of d electrons.

The instability of the antiferromagnetic state in band magnets at strong magnetic fields was predicted by Wohlfarth as long ago as 1963 [14]. Wohlfarth showed that under certain conditions, an antiferromagnetic metal may undergo a phase transition into a ferromagnetic state at temperatures close to absolute zero in a strong external magnetic field $H = H_{cr}$. Wohlfarth's work [14] implied that if the coefficients $A < 0$ and $B < 0$ in the Landau expansion for the free energy,

$$F = \frac{1}{2}A\sigma^2 + \frac{1}{2}B\sigma^4 - H\sigma, \qquad (5)$$

the antiferromagnetic state would be unstable, and a phase transition into the ferromagnetic state would occur at some critical value of the field depending on the ratio of these coefficients:

$$H_{cr} = 2/3(A^3/3B)^{1/2}. \qquad (6)$$

Despite the fact that the energy difference between the ferro- and antiferromagnetic states in Pt_3Fe is small, in our experiments, we were most likely only able to observe the onset of this process in pulsed fields of order $200\,kOe$ (see Fig. 3).

A much more complex situation arises in alloys of non-stoichiometric composition because of the formation of induced magnetic moments on the paramagnetic platinum atoms (band electron metamagnetism). This phenomenon was also predicted by Wohlfarth [15].

Magnetic moments can only arise in the case where the total internal energy at $T = 0$ as a function of the relative magnetization ς has two minima: one at $\varsigma = 0$, and one at $\varsigma < 1$. In addition, this effect can only be observed in materials for which the state density curve has a certain shape: $N(\epsilon)$ must have positive curvature, and the Fermi level must be located near the peak in $N(\epsilon)$ [15].

It is currently known [16] that the induced magnetic moments may be due not only to strong external fields but also to the internal (exchange) fields caused by the exchange interactions between the d band electrons and the localized moments of the nearest-neighbor atoms; what is important from the point of view of the induced magnetic moment is the size of the total (external plus internal) effective field acting on the atom.

The mean magnetic moments induced on the platinum atoms in the iron–platinum alloys by the effective fields can be calculated from the size of the discontinuity in the magnetization at the critical field value at 4.2 K. These data are as follows:

c_{Fe}, at%	25.2	27	28	30.5
μ_{Pt}, μ_B	0.1	0.1	0.14	0.28

(The "additional" iron atoms are also assumed to contribute to the discontinuity in magnetization.)

These data imply that the magnetic moments are in reasonable agreement with the values $\mu_{Pt} = 0.25–0.3\mu_B$ obtained for ordered and disordered ferromagnetic iron–platinum alloys in neutron diffraction experiments.

2. INSTABILITY OF THE MAGNETIC STATES
IN IRON–PLATINUM AND IRON–RHODIUM
ALLOYS TO ISOTROPIC COMPRESSION

2.1. ANOMALIES IN THE PHYSICAL PROPERTIES OF
Fe–Rh ALLOYS IN THE PHASE TRANSITION REGION

Along with the technical difficulties associated with research at high pressures, there are also some physical limitations on the applicability of certain methods. We therefore studied a variety of physical properties (electrical resistance, thermoelectromotive force, differential magnetic susceptibility, and heat capacity); this allowed us to identify the anomalies in the vicinity of the phase transitions over a wide temperature range from 4.2 to 1000 K. The phase transition temperatures obtained from the measurements of the various physical quantities as a function of temperature were in good agreement, even though they did not have completely identical values.

The specific electrical resistance as a function of temperature is shown for alloys of various composition at high temperatures in Fig. 6. As one can see from the figure, the resistance increases until the temperature T_{cr}, falls sharply in the vicinity of the phase transition, and increases with temperature once again in the ferromagnetic region. The electrical resistance experiences a weak break at the Curie point. Significant temperature hysteresis is observed as the samples are heated and cooled in the vicinity of the AFM–FM phase transition. The hysteresis varies from 8 to 15 K, and increases as the iron content of the alloy increases. The presence of hysteresis is a distinguishing feature of a first-order phase transition. Figure 6 indicates that the phase transition in the alloy with $c_{Fe} = 49.5$ at% is much sharper than in the alloy with 50.5 at% Fe.

It should be noted that a slow, linear decrease in the electrical resistance with temperature is observed for all of the alloys in the paramagnetic region, i. e., the temperature dependence of the resistance is similar to that of a semiconductor. A similar $\rho(T)$ function was observed for iron–rhodium alloys in [17], where this fact was used in support of the idea that iron–rhodium alloys of quasi-equiatomic composition were semimetals.

As is well known, the anomalies and other kinetic phenomena are due to the scattering of conduction electrons near the Curie and Néel points. The theoretical calculations carried out in [18, 19], where the presence of gaps in the conduction electron spectrum due to the magnetic lattice doubling showed that the temperature dependence of the thermoelectromotive force in the vicinity of the Curie and Néel points is determined by the ratio of the vectors $\mathbf{K}_A$ and $\mathbf{K}_F$ (where $\mathbf{K}_F$ is the Fermi quasimomentum and $\mathbf{K}_A$ is the antiferromagnetism vector in the inverse lattice). If $K_A \approx K_F$, the temperature dependence of the thermoelectromotive force is determined by critical scattering on fluctuations in the large-scale

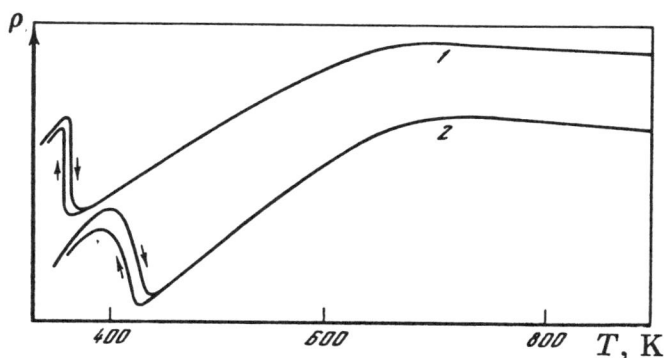

Figure 6 Electrical resistance of iron–rhodium alloys as a function of temperature. (1) $Fe_{49.5}Rh_{50.5}$; (2) $Fe_{50.5}Rh_{49.5}$.

order, and has a "maximum-type" anomaly; if $K_A \ll K_F$, the main contribution to the critical anomaly comes from scattering on small-scale fluctuations (in ferromagnetic metals, this occurs for $K_A \sim 0$), and a "jump discontinuity-type" ferromagnetic anomaly should be observed at the Néel point.

The experimental data indicate that the anomalies in the temperature dependences of the thermoelectromotive force and electrical resistance of the alloys in

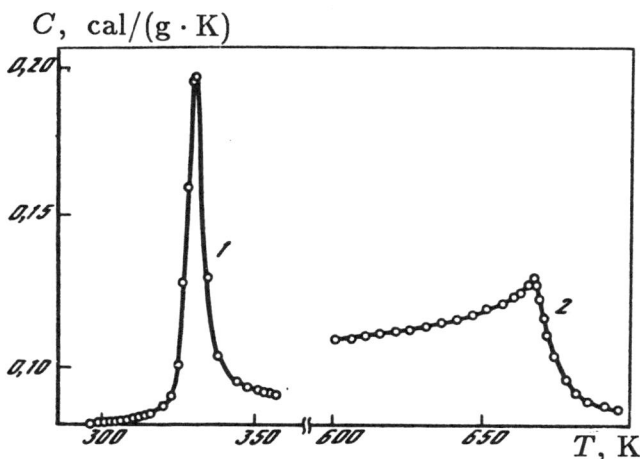

Figure 7 Heat capacity of the alloy $Fe_{49.5}Rh_{50.5}$ as a function of temperature near the phase transitions. (1) AFM–FM; (2) FM–PM.

the first-order phase transition are similar. However, as will be shown below, decreasing the interatomic distances in the alloy affects the anomalies in the electrical resistance and thermoelectromotive force differently.

According to the Ehrenfest equations, a phase transition can be characterized by the orders of the derivatives of the thermodynamic potential which undergo finite changes during the phase transition: the first derivatives of the thermodynamic potential with respect to temperature and pressure (volume, entropy, etc.) have jump discontinuities in first-order phase transitions, while the second derivatives (heat capacity, compressibility, etc.) have jump discontinuities in second-order phase transitions.

Measurements of the heat capacity of iron-rhodium alloys with equiatomic composition near the phase transitions showed that the anomalies observed in the heat capacity in the first- and second-order phase transitions are appropriate to these

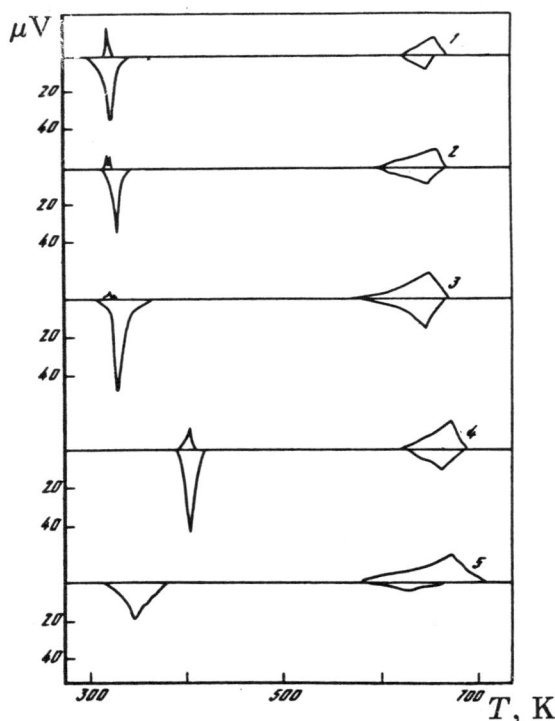

Figure 8   Thermograms of iron–rhodium alloys with various compositions. (1) $Fe_{49.0}Rh_{51.0}$; (2) $Fe_{49.5}Rh_{50.5}$; (3) $Fe_{50.0}Rh_{50.0}$; (4) $Fe_{50.5}Rh_{50.0}$; (5) $Fe_{51.0}Rh_{49.0}$.

phase transitions (Fig. 7). The absolute size of the anomaly in the AFM–FM phase transition turned out to be 5–6 times larger than that in the FM–PM phase transition.

The method of differential thermal analysis (DTA), which essentially reduces to determining the endothermicity or exothermicity of reactions in a material by observing the way in which the temperature of a sample varies with time, has come into wide use for studying first-order phase transitions.

We used DTA to study iron–rhodium alloys in the vicinity of the phase transitions. The thermograms (which were obtained in the traditional way) are shown in Fig. 8. As one can see from the figure, the anomalies in the thermograms differ somewhat as a function of alloy composition, which is indirect evidence that the samples have varying degrees of homogeneity. The thermograms show clear anomalies around the ferromagnetic-paramagnetic second-order phase transition, where (as is well known) heat is neither released nor absorbed. This effect is due to the fact that the function $\Delta T/T$ is measured in a dynamic mode, i. e., $dT/dt \neq 0$ (where t is time).

At high pressure, we used a modified differential thermal analysis method with a non-homogeneous temperature field [20]; this modified method has certain advantages over the traditional method. In measurements using this method, a characteristic anomaly (with two extrema) which differs substantially from the thermograms obtained in the usual way occurs in the vicinity of first-order phase transitions. Each pair of opposing maxima is a result of the passage of the internal boundary between phases (which is normal to the temperature gradient) past each of the thermocouples in turn. The sizes and shapes of the maxima are determined by various factors, including the temperature gradient, the rate of change in the temperature, and the heat capacity and thermal conductivity of the sample. Thus, a fairly large difference between the coefficients of thermal conductivity in the two phases would mean that the first maximum might not be identical to the second. This indeed explains some of the difference between the ΔT values before and after the phase transition.

2.2. THE EFFECT OF ISOTROPIC PRESSURE ON THE PHYSICAL PROPERTIES OF IRON–RHODIUM ALLOYS OF EQUIATOMIC COMPOSITION. THE p–T MAGNETIC PHASE DIAGRAM

Measurements of the electrical resistance, differential magnetic susceptibility, and thermoelectromotive force were carried out at various fixed pressures in order to determine the effects of pressure on the AFM–FM and FM–PM phase transitions and construct p–T magnetic phase diagrams for the alloys.

The high pressures were obtained using the "Toroid" pressure chambers described in [21], which enabled us to obtain pressures of up to 100 kbar. An internal heating element enabled the sample to be heated to 1000 K. The samples under study were placed in an ampule filled with a hydrostatic fluid. In addition to the sample, the ampule also contained a heating element, pressure sensors, thermocouples, and measuring coils. Up to 18 electrical wires could be fed into the ampule; this, together with the hydrostatic conditions provided much room for experimentation. The "Toroid" pressure chambers are distinguished by simplicity and reliability of operation, as well as high reproducibility of their results.

Fig. 9 shows the electrical resistance isobars for an alloy of equiatomic composition. For relatively low pressures (up to 40–50 kbar), the temperature dependence of the electrical resistance is similar to that observed at atmospheric pressure (see Fig. 6). Both transitions are fairly sharply localized. Increased pressure produces a decrease in electrical resistance in the antiferro- and paramagnetic phases, but has practically no effect on the resistance in the ferromagnetic phase. The AFM–FM transition becomes less sharp and the amount of temperature hysteresis decreases under pressure. The FM–PM transition becomes unresolvable at pressures $p > 50$ kbar. The trend is, however, clear: T_{cr} increases with increasing pressure and the Curie temperature (T_C) decreases, which suggests that we are near a triple point. The coordinates of the triple point cannot be determined from the electrical resistance measurements, since the smallest difference observed was $T_{cr} - T_C \approx 50$ K, but we can infer that the ferromagnetic phase vanishes completely for pressures above 70 kbar.

This conclusion is supported by the susceptibility curves as a function of temperature at high pressure shown in Fig. 10. As one can see from the figure, the region over which the high-susceptibility ferromagnetic phase exists gets smaller; at $p > 65$ kbar, only the weak breaks in the $\chi(T)$ curves due to the AFM–PM phase transition remains. It turned out that the $\chi(T)$ curves (which have large anomalies both in the vicinity of the second-order FM–PM phase transition and in the vicinity of the first-order AFM–FM phase transition) could be used to more precisely determine the temperatures of the phase transitions $(T_{cr}$ and $T_C)$ at high pressures $(p > 50$ kbar).

The thermoelectromotive force is shown in Fig. 11 as a function of temperature for various pressures. The anomalies in the thermoelectromotive force and electrical resistance during the AFM–FM phase transition are quite different, and show quite different behavior with increasing pressure. For example, the size of the discontinuity in electrical resistance is a weak function of pressure, and high pressure suppresses the discontinuity in the thermoelectromotive force at T_{cr}. The thermoelectromotive force as a function of temperature is affected in the paramagnetic region: the slope $\alpha(T)$ changes from positive to negative for pressures greater than p_{tr} (p_{tr} is the triple point pressure). On the other hand, the coefficient of electrical resistance is practically independent of pressure. The effects of pressure on the alloy electron energy spectrum evidently affect the various kinetic coefficients differently.

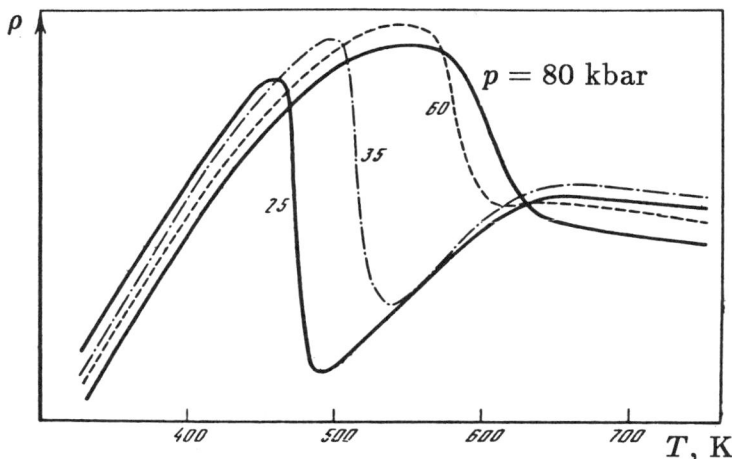

Figure 9   Isobars of electrical resistance for the iron–rhodium alloy $Fe_{49.5}Rh_{50.5}$.

Figure 12 shows the p–T magnetic phase diagram for iron–rhodium alloys of equiatomic composition. This diagram proves the existence of the triple point in the p–T plane where the antiferro-, ferro- and paramegnetic phases coexist, as predicted by Ponyatovskii and coworkers [22]. Direct phase transitions from the antiferromagnetic state into the paramagnetic state can occur above the triple point. The experimentally determined coordinates of the triple point are: $T_{tr} = 593$ K and $p_{tr} = 59$ kbar.

For pressures between 0 and 30 kbar, the temperature of the phase transition, T_{cr}, is a linear function of pressure with slope 5.38 K/kbar. This value is in good agreement with the results in [23], where a value of 5.75 K/kbar was obtained under hydrostatic pressure, but differs from the results in [24] (4.3 K/kbar) and [25]. A linear dependence of T_{cr} on pressure was predicted in the theoretical work of Grazhdankina, et al. [26], where the effect of pressure on phase stability in cubic crystals during antiferromagnetic-ferromagnetic phase transitions was studied. The value of $dT_{cr}/dp = 2.9$ K/kbar obtained in this paper is, however, quite different from our results.

As first noted in [27], dT_{cr}/dp is observed to start decreasing (i. e., the function $T_{cr}(p)$ deviates from linearity) at higher pressures ($p > 30$ kbar), and $dt_{cr}/dp = 1.9$ K/kbar at a pressure of approximately 60 kbar. The existence of a nonlinear relationship between the temperature of the antiferromagnetic–ferromagnetic phase transition and pressure was predicted theoretically in [28]: $T_{cr}(p) = T_0 + k\sqrt{p}$, where T_0 is the temperature of the phase transition at atmospheric pressure, and k is the modulus of hydrostatic compression. Substituting our experimental values into the formula, we estimate a value of $k = 34\,\text{K/kbar}^{1/2}$ for the constant, and a quite similar value ($k = 29\,\text{K/kbar}^{1/2}$) results from the expression $dT_{cr}/dp = k/2\sqrt{p}$, which is an indication of the applicability of the model proposed by Petrov.

As one can see from the phase diagram, the Curie temperature of the alloy falls off almost linearly with increasing temperature, and our results are not at variance with the data in the literature: Dubovka [25] quotes a value of $dT_C/dp = -0.9\,\text{K/kbar}$. Above the triple point, the temperature of the antiferromagnetic–ferromagnetic transition temperature (T_N) increases linearly with pressure up to 100 kbar. The linear dependences of T_C and T_N on pressure can be effectively described within the framework of the s–d exchange model [28].

Since our experiments imply the existence of a triple point in the p–T phase diagram, we needed to resolve the questions surrounding the nature of the AFM–PM phase transition, since there is a contradiction between the theoretical predictions [29] and the experimental results [27]. We chose two criteria for first-order phase transitions on the basis of the capabilities of experimental measurement techniques at high pressures: the presence of temperature hysteresis during the phase transition and the release or absorption of heat. In agreement with results in the literature, our measurements showed a significant amount of temperature hysteresis in the temperature dependences of the susceptibility and other variables during the antiferromagnetic–ferromagnetic phase transition, characterizing it as a first-order transition. The amount of hysteresis is a strong function of the amount of pressure applied (see inset in Fig. 12). The figure indicates that the amount of hysteresis decreases with pressure, and is equal to zero above the triple point to within the experimental errors.

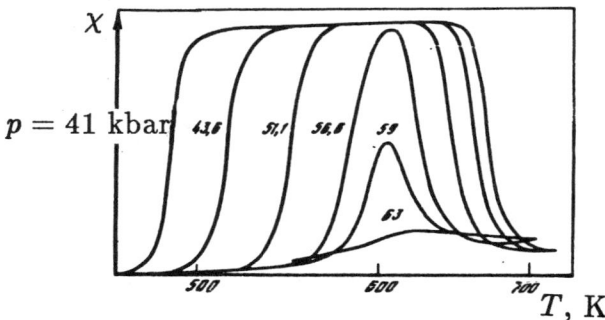

Figure 10   Susceptibility as a function of temperature for the alloy $Fe_{49.5}Rh_{50.5}$ at various pressures.

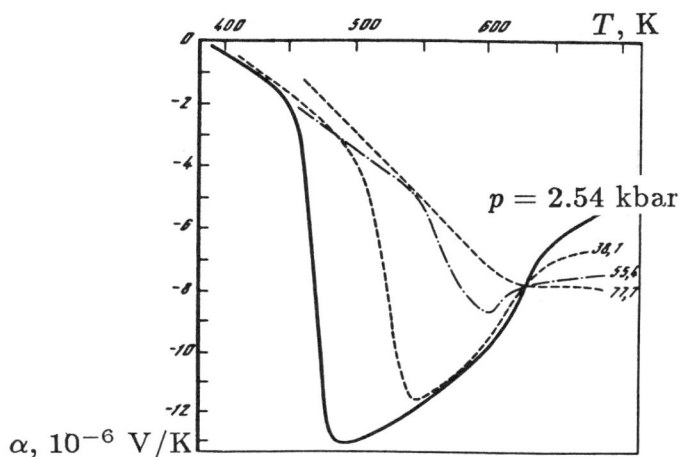

Figure 11   Thermoelectromotive force isobars for the alloy $Fe_{49.5}Rh_{50.5}$.

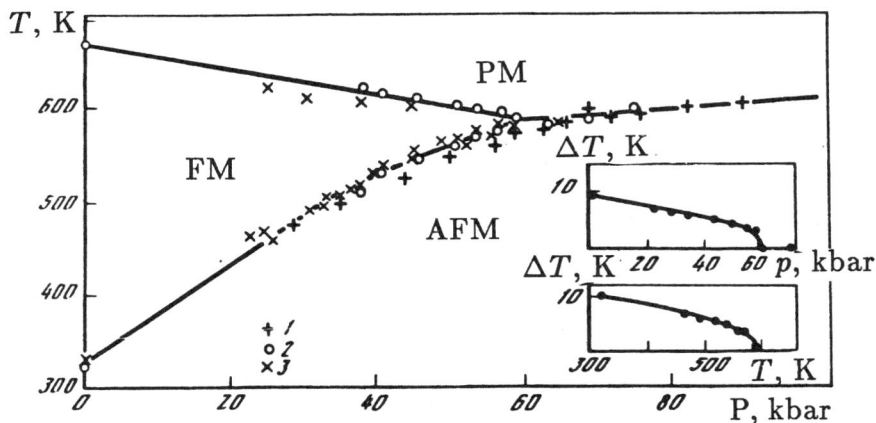

Figure 12   Magnetic p–T phase diagram for an equiatomic iron–rhodium alloy constructed from the results of (1) differential thermal analysis, susceptibility measurements (2), and electrical resistance measurements (3). Inset: Hysteresis ΔT as a function of temperature during the AFM–FM phase transition.

This provided a basis for the conjecture that the antiferromagnetic–paramagnetic phase transition above the triple point is a second-order phase transition;

this is in agreement with the calculations of Hargitai, but not with the experimental data of Leger, et al. [27]. In this context, we carried out careful differential thermal analysis studies of the alloys at pressures above and below p_{tr}. The results of these experiments are shown in Fig. 13. We shall discuss some of the characteristics of the thermograms shown in the figure in some detail. At pressures below p_{tr}, the anomalies in the thermograms in the vicinity of the antiferromagnetic–ferromagnetic phase transition at different pressures are similar. Some difference in the absolute magnitude of the signal is natural, since it depends both on the rate of change of the temperature and the parameters of the medium transmitting the pressure. In addition to the anomalies associated with the AFM–FM phase transition, the figure also shows some of the anomalies associated with the FM–PM phase transition (i. e., a second-order phase transition) for 42.2 and 53 kbar pressure. However, the size of the measured anomaly

in the second-order phase transition is several times smaller than that for the first-order phase transition, which corresponds to measurements at atmospheric

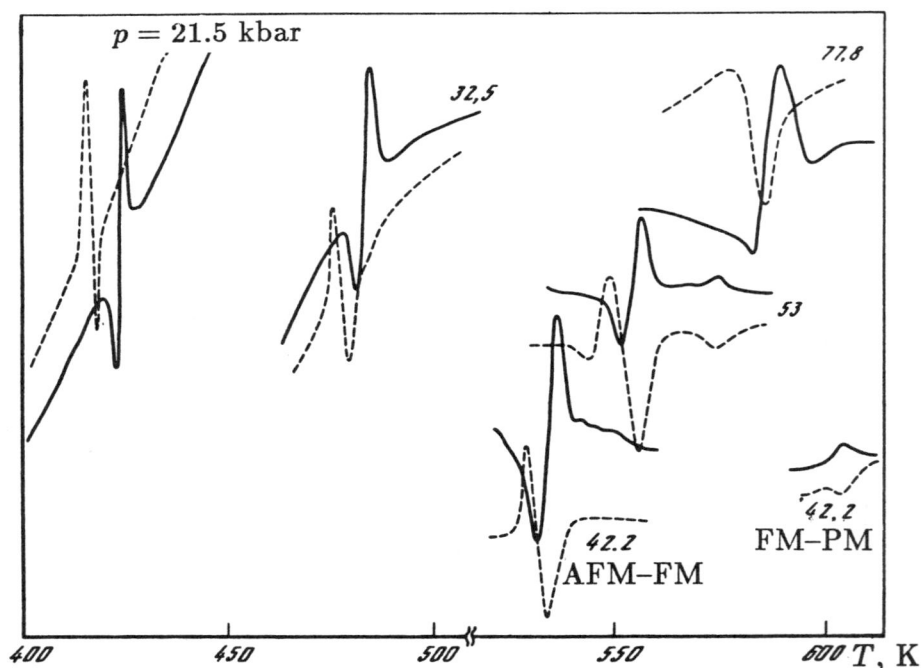

Figure 13   Thermograms of the iron–rhodium alloy $Fe_{49.5}Rh_{50.5}$ at various pressures.

pressure which were made in a vacuum.

The shapes of the anomalies associated with the AFM–PM phase transition at pressures above p_{tr} are different from those of the anomalies associated with the AFM–FM phase transition. Since the physical properties of the alloy (which determine the shapes of the anomalies associated with the phase transition) in the ferromagnetic and paramagnetic states are different, this change in signal shape is natural. This can be interpreted physically as a change in the boundary between the phases, i. e., below p_{tr}, the interphase boundary separates ferromagnetic ordering from antiferromagnetic ordering, while it separated antiferromagnetic ordering from ferromagnetic ordering above this pressure. At $p > p_{tr}$, the anomalies are closer in size to those associated with the AFM–FM phase transition. It is thus impossible to reach an unambiguous conclusion about the nature of the AFM–PM transition from our experiments. It appears that only neutron diffraction studies at high pressure will resolve this problem.

2.3. THE EFFECT OF PRESSURE ON THE PROPERTIES OF FERROMAGNETIC IRON–RHODIUM ALLOYS

The numerous experimental and theoretical papers available on iron–rhodium alloys have generally discussed AFM–FM phase transitions in alloys of quasi-equiatomic composition. The effects of pressure on the properties of ferromagnetic alloys of nonstoichiometric composition were not studied. Since the antiferromagnetic state in alloys of stoichiometric composition becomes more stable with decreasing interatomic separation, one might suppose that an analogous situation would hold for the effects of high pressure on ferromagnetic iron–rhodium alloys. Moreover, the alloys with iron concentrations of 51.1 and 52.0 at% Fe that we studied are on the boundary of the transition region between the antiferromagnetic and ferromagnetic phases, and theoretical calculations [30, 31] indicate that the difference between the energy states in the antiferro- and ferromagnetic phases is small.

Figure 14 shows the relative electrical resistance of three alloys with iron concentrations of 50.0, 51.5, and 52.0 at%, respectively, measured at room temperature as a function of pressure. As one can see from the figure, pressure has a different effect on antiferro- and ferromagnetic alloys. The electrical resistance is lower in the antiferromagnetic state, and is only a slight function of pressure. No anomalies were observed. In the ferromagnetic alloys, a sharp increase in resistance (similar to that observed in first-order antiferromagnetic–ferromagnetic phase transitions in stoichiometric-composition alloys) was observed at pressures between 25 and 45 kbar. It can be shown that the $c_{Fe} = 51.5$ at% alloy undergoes a pressure-induced FM–AFM phase transition. Under further increases in pressure, the electrical resistance of the alloy increases, and the behavior of the electrical resistance as a function of pressure is similar to that observed for alloys

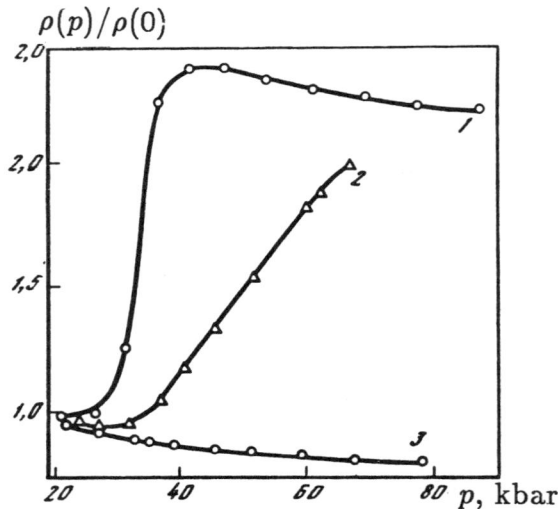

Figure 14   Electrical resistance of iron–rhodium alloys of various composition as a function of pressure at room temperature. (1) $Fe_{51.5}Rh_{48.5}$; (2) $Fe_{52.0}Rh_{48.0}$; (3) $Fe_{50.0}Rh_{50.0}$.

in the antiferromagnetic state. In the alloys with iron concentration 52 at%, the electrical resistance increases more gradually with increasing pressure, and the phase transition occurs at pressures above 50 kbar.

Since the pressure-induced FM–AFM transition is sharper in the alloy with $c_{Fe} = 51.5$ at%, we used this alloy for a detailed study of the electrical resistance and susceptibility as a function of temperature at various pressures.

The isobars of electrical resistance for the alloy are shown in Fig. 15. At low pressures (up to 25 kbar), there is one anomaly in the function $\rho(T)$, at the Curie temperature. A discontinuity appears in the $\rho(T)$ curves at pressures of between 25 and 50 kbar, and it increases in size as the pressure increases. This anomaly is due to the AFM–FM phase transition. Since Figure 14 implies that the FM–AFM phase transition occurs on the pressure interval from 25 to 45 kbar, we can conjecture that the ferro- and antiferromagnetic phases coexist over a fairly broad range of temperatures and pressures, and that the size of the discontinuity in the electrical resistance depends on the volume occupied by the antiferromagnetic phase. The presence of anomalously large temperature hysteresis on this pressure interval in the vicinity of the AFM–FM phase transition is confirmation of the fact that the magnetic structure is not homogeneous.

For pressures greater then 50 kbar, the size of the discontinuity in electrical resistance first stabilizes and then begins to decrease with increasing pressure.

The temperature hysteresis in the vicinity of the phase transition also decreases with increasing pressure. At these pressures, the entire sample has apparently undergone the phase transition into the antiferromagnetic state.

The susceptibilities $\chi(T)$ as a function of temperature at various pressures were studied at low temperatures in Itskevich fixed-pressure chambers [32], which enabled us to obtain pressures of 12 kbar at room temperature and 10.5 kbar at $T = 77$ K.

The functions $\chi(T)$ for the alloy with $c_{Fe} = 51.5$ at% confirm that a FM–AFM

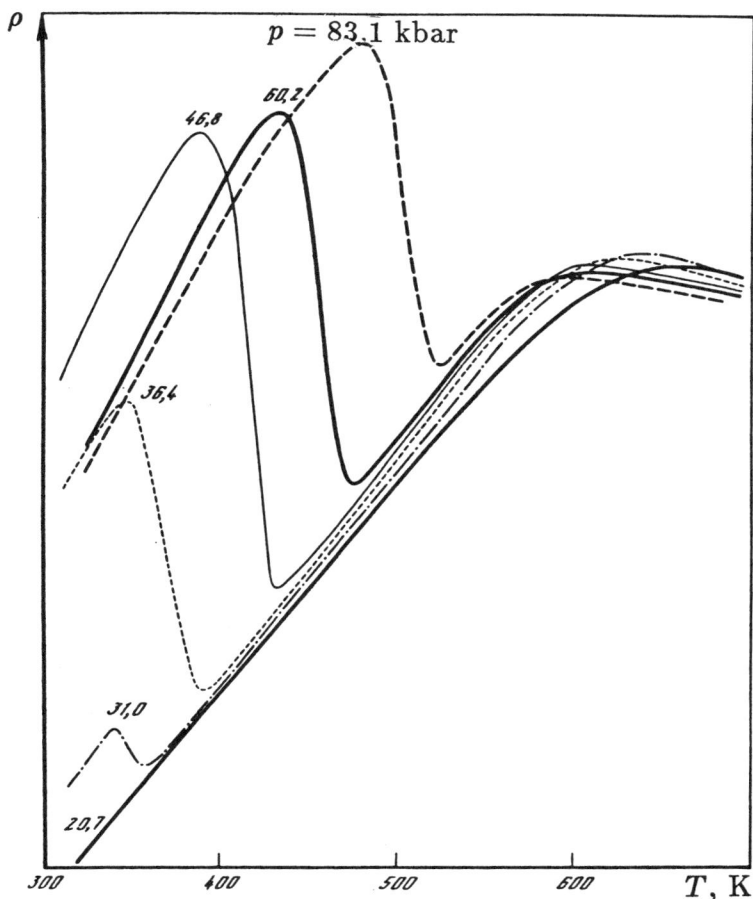

Figure 15   Electrical resistance of the ferromagnetic alloy $Fe_{51.5}Rh_{48.5}$ at various pressures.

transition does occur. However, since the curves $\chi(T)$ are sections of hysteresis loops (like those that might be obtained from studying the functions $\chi(H)$ for incomplete magnetization cycles [33]), the temperature of the phase transition cannot be determined from the graphs. The experiments imply that the hysterisis becomes much larger at these pressures, reaching 100–200 K; in addition, signs of antiferromagnetic order are already present at 77 K by approximately 5 kbar pressure.

Fig. 16 shows the magnetic phase diagram of the alloy with $c_{Fe} = 51.5$ at% constructed on the basis of our experiments. It should be noted that since the FM–AFM phase transitions occur over a very wide range of temperatures and pressures, these phase transitions would be better described in terms of a phase transition region, rather than a phase boundary. We can say something definite about the Curie temperature as a function of pressure, which is analogous to that observed for the equiatomic-composition alloy: the value of $dT_C/dp = -0.9$ K/kbar for the alloy with 51.5 at% Fe is similar to the value of $dT_C/dp = -1.2$ K/kbar for the alloy with $c_{Fe} = 50.0$ at%.

On the basis of the experimental electrical resistance and susceptibility functions for the alloy with $c_{Fe} = 51.5$ at%, we can infer the following facts about the properties of the phase transitions at high pressure. The region with ferromagnetic ordering becomes narrower with increasing pressure. For pressures $p > 50$ kbar, the points shown in the phase diagram correspond to the center of the phase transition region. For pressures $p < 50$ kbar, the temperatures of the phase transitions were determined for samples that originally consisted of two phases, so that the points on the figure do not correspond to the center of the phase transition region, and they provide only approximate information on the temperature hysteresis. The points on the phase diagram at room temperature show the center and edges of the phase transition region at this temperature. The low-temperature portion of the diagram has been drawn in on a provisional basis, since it was constructed using three experimental facts: 1) the analysis showed that this alloy was ferromagnetic all the way down to liquid-helium temperatures at atmospheric pressure; 2) the isobars for the susceptibility as a function of temperature at low temperature indicate that the edge of the phase transition region passes close to the point with coordinates $p = 5$ kbar, $T = 77$ K; and 3) the phase transition region and hysteresis region become narrower with increasing pressure. On the basis of these facts, we conjecture that there exists a range of pressures at $T = 0$ where the AFM and FM phases coexist. Moreover, there exists a temperature T_x below which the alloy may have a metastable state in which the antiferro- and ferromagnetic states coexist at $p = 0$.

Even pressures up to 100 kbar turned out not to be high enough to completely suppress the ferromagnetic phase in this alloy. However, the shape of the phase diagram, the nature of the anomalies in electrical resistance in the phase transition region, and the decrease in hysteresis with increasing pressure lead us to

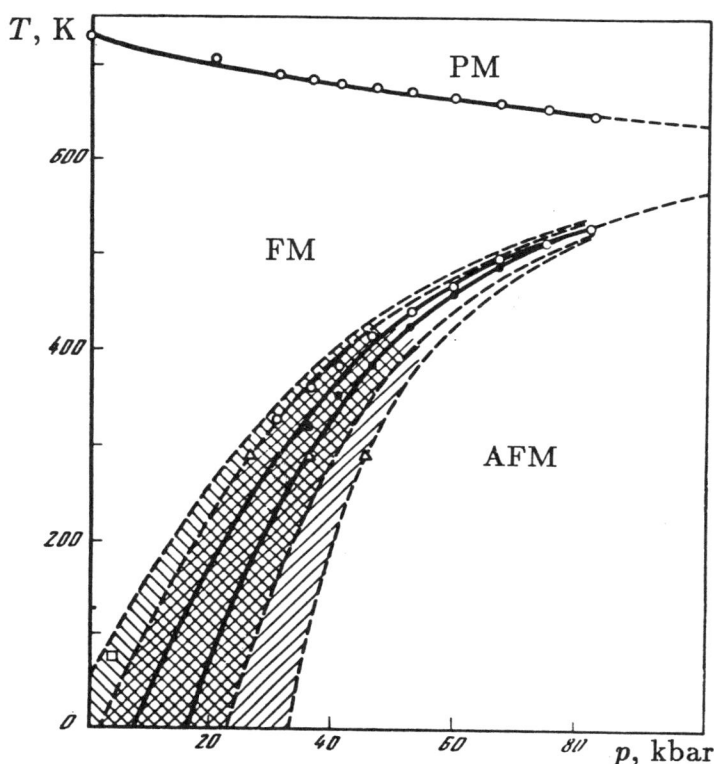

Figure 16   Magnetic p–T phase diagram for the alloy $Fe_{51.5}Rh_{48.5}$.

conjecture that the antiferromagnetic order becomes more and more preferable with increasing pressure, and at some pressure, the alloy will become antiferromagnetic over the entire range of temperatures on which magnetic ordering is observed. The quantity $dT_{cr}/dp = 2.2$ K/kbar at pressures of 70–80 kbar, which is not much different from the value for the stoichiometric-composition alloy. Approximation of the phase boundaries at higher pressures yields the following triple point coordinates: $p \approx 130$ kbar and $T \approx 600$ K.

2.4. EFFECT OF ISOTROPIC PRESSURE ON THE MAGNETIC PROPERTIES OF Pt–Fe ALLOYS

We studied the effect of hydrostatic pressures up to 40 kbar on the magnetic phase transitions in iron–platinum alloys with compositions ranging from 24.8

to 32.2 at% Fe. The phase transition temperatures for the alloys with iron concentration less than 30 at% were determined from the anomalies on the curves for the electrical resistance as a function of temperature, while those for the alloys with iron concentration greater than 30 at% were determined from the anomalies in the differential magnetic susceptibility as a function of temperature.

The shifts in the magnetic phase transition temperatures T_1 and T_2 for the

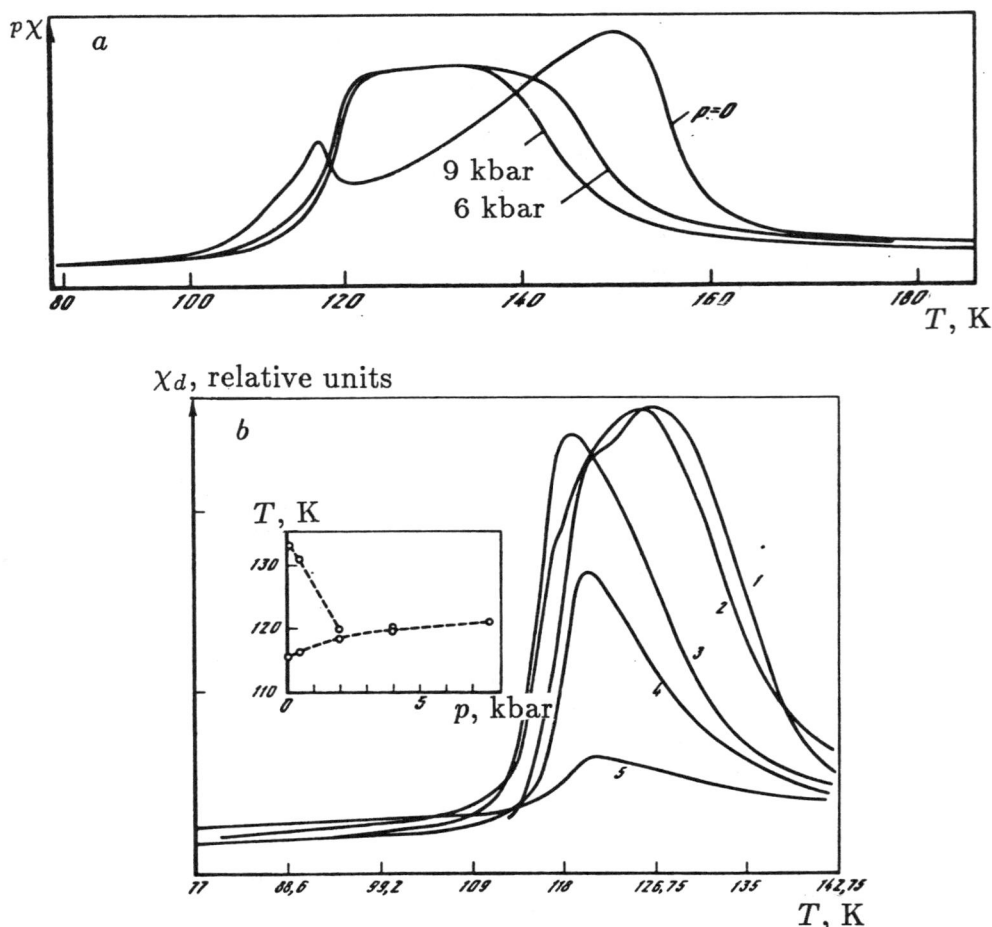

Figure 17   Differential magnetic susceptibilities under pressure for iron–rhodium alloys as a function of temperature. (a) 32.2 at% Fe; (b) 30.65 at% Fe ((1) 1 bar, (2) 0.34, (3) 1.92, (4) 3.92, (5) 7.58 kbar); Inset: magnetic phase diagram for a 30.65 at% Fe iron–rhodium alloy.

alloys with $c_{Fe} \leq 28\,at\%$ over the range of pressures studied turned out to be very small–$\lesssim 1\,K$, i. e., no greater than the errors in determining these temperatures. In the alloys with $c_{Fe} \geq 30.5\,at\%$, on the other hand, the pressure has a substantial effect on the phase transition temperature.

Figure 17a shows the differential magnetic susceptibility of the alloy with 32.2 at% Fe (which has ferromagnetic order at room temperature) over a fairly wide range of temperatures $T_1 - T_2$.

As the figure indicates, the ferromagnetic region becomes noticeably narrower with increasing pressure, and the two peaks in the $\chi_d(T)$ curve are replaced by a plateau (in which case the magnetic phase transition temperature was determined from the maximum value of the derivative $d\chi/dt$). The measured results for this alloy are given in Table 3.

Linear extrapolation of the functions $T_1(p)$ and $T_2(p)$ to higher pressures yields the following values for the critical point at which the ferromagnetic state vanishes: $p_{tr} = 21\,kbar$, $T_{tr} = 125\,K$. Measurements carried out at higher pressures confirmed that there was only a single transition at temperature T^*, with $dT^*/dp \approx 0.15\,K/kbar$.

The fact that the p–T phase diagram has a triple point was confirmed by studies of the alloy with 30.65 at% Fe, in which the ferromagnetic region is narrower (Fig. 17b). Although the two phase transitions can still be resolved in Curves 1 and 2, which correspond to pressures of $p = 0$ and $p = 0.34\,kbar$, respectively (the phase transitions become more resolvable when a weak constant magnetic field is imposed), they start to become non-resolvable and the susceptibility peak becomes much smaller at pressures $p \sim 2\,kbar$. Curve 5 $(p = 7.6\,kbar)$

Table 3. Effect of Pressure on Magnetic Phase Transition Temperatures in the Alloy $Fe_{32.2}Pt_{67.8}$

p, kbar	T, K	ΔT, K	$\frac{dT}{dp}, 10^{-3}\,K/bar$	$\frac{1}{p}\frac{dT}{dp}, 10^{-5}\,bar^{-1}$
0	117			
5.6	119.25	2.25	0.40	0.34
8.5	120	3	0.35	0.30
		Mean	0.375	0.32
0	155			
6.0	146.25	- 8.75	-1.46	-0.94
9.0	142.75	-12.25	-1.36	-0.88
		Mean	-1.41	-0.91

is of the same form as in the alloy with $c_{Fe} = 30.5$ at%, where only a single transition from the antiferromagnetic state into the paramagnetic state is observed.

The phase diagram for this alloy is shown in the inset to Fig. 17b.

Thus, hydrostatic pressure suppresses ferromagnetism in Fe–Rh, Fe–Rh–Ir [34], and other alloys. Like the Pt–Fe alloys with high iron concentrations ($c_{Fe} \gtrsim 30.5$ at%), many band magnetic substances are characterized by large derivatives dT/dp.

3.   ELECTRONIC BAND STRUCTURE OF Pt–Fe AND Fe–Rh ALLOYS

It is natural to assume that the observed instability of the magnetic states in Pt–Fe and Fe–Rh alloys with respect to alloy composition and external effects is due to the electronic structure of these alloys. For example, according to [16], the phenomenon of band electron magnetism causes the density of states to depend on energy to some extent.

We therefore carried out a self-consistent calculation of the band structures, density of states, and Fermi surfaces in the paramagnetic state of the completely ordered stoichiometric-composition alloys Pt_3Fe and FeRh using the linearized Korringa–Kohn–Rostocker method [36].

3.1. IRON–PLATINUM ALLOYS

Figure 18a shows the density of states in Pt_3Fe calculated by the tetrahedral method at 165 points in the 1/48 irreducible part of the Brillouin zone. The figure implies, in agreement with the predictions of Wohlfarth, that the Fermi level in Pt_3Fe is indeed located near a large peak in the density of states curve. The density of states in the Fermi level turned out to be equal to 81.8 states/(Ryd · elem. cell). Figure 19a shows cross sections through the Fermi surface on high-symmetry planes in the paramagnetic phase.

From the neutron diffraction measurements, we know that the elementary magnetic cell is doubled during the transition to antiferromagnetic ordering, and becomes tetragonal base-centered. The Brillouin zone also changes; the boundaries of the new Brillouin zone are shown by the dot-dash lines in Fig. 19a. The new Brillouin zone boundaries pass through the inverse lattice vectors joining points Γ and M and R and X, and all of the states lying outside these boundaries must be balanced within the reconstructed Brillouin zone. The smallest "electron sphere" is almost identical in size and shape to the largest hole pocket centered on the point Γ, while the "electron ellipsoid" centered on the point X is nearly identical to the hole pocket at point R.

N, states/(Ryd · elem. cell)

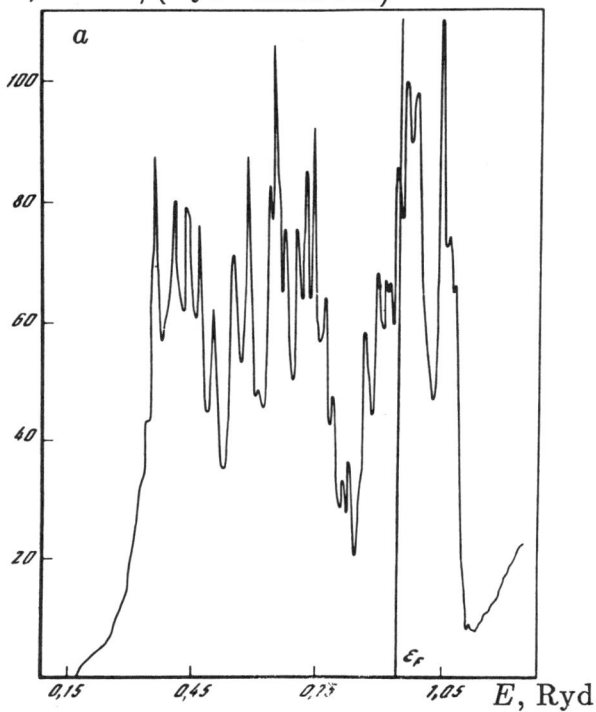

N, states/(Ryd · elem. cell)

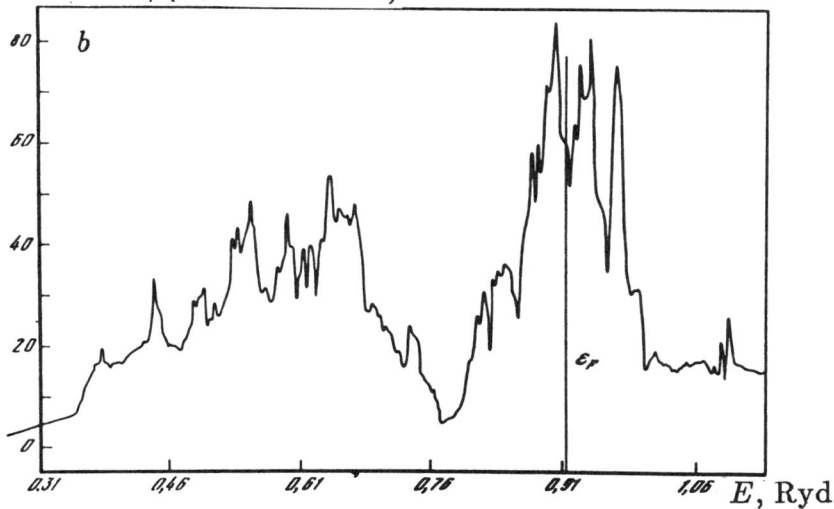

Figure 18 Density of states for ordered alloys in the paramagnetic state. (a) Pt₃Fe; (b) FeRh.

Thus, the Fermi surface has identical electron and hole surfaces shifted by the vector $\mathbf{Q}_1 = \pi/a[110]$, i. e., we have a situation similar to that observed in chromium in antiferromagnetically-ordered Pt_3Fe—the formation of triplet electron–pair holes [37]. In this case, the vector $\mathbf{Q}_1$ is identical to the spin density wave (SDW) vector, which will have nodes precisely at the points where the platinum atoms are located; this confirms the absence of magnetic moments on the latter in the "ideal" magnetic structure.

The electron-hole pairing results in a decrease in the density of states on the Fermi surface during the phase transition into the antiferromagnetic state, which is in agreement with the results of measurements in the antiferromagnetic state and calculations for the paramagnetic state: $\gamma = 5.5$ [38] and $14.2\,\mathrm{mJ}/(\mathrm{mole} \cdot \mathrm{K}^2)$, respectively.

The mean number of electrons per atom decreases with increasing iron concentration, which should make the conditions for overlap between the shifted electron and hole portions of the Fermi surface worse and lead to suppression of the state with wave vector $\mathbf{k}_1 = [1/2\ 1/2\ 0]$. At the same time, the onset of antiferromagnetic structure with $\mathbf{k}_2 = [1/2\ 0\ 0]$ is evidently the result of the "overlap" between the electron ellipsoid at point X and the intermediate hole pocket at the point Γ upon translation by the vector $\mathbf{Q}_2 = \pi/a[100]$. The function $N(\epsilon_F)$ should decrease even further in the region where the two magnetic reflections with $\mathbf{k} = [1/2\ 1/2\ 0]$ and $[1/2\ 0\ 0]$ coexist, which is also in agreement with the experimental results: $\gamma = 2.7\,\mathrm{mJ}/(\mathrm{mole} \cdot \mathrm{K}^2)$.

A calculation of the static susceptibility function $\chi^0(\mathbf{q})$ carried out in the

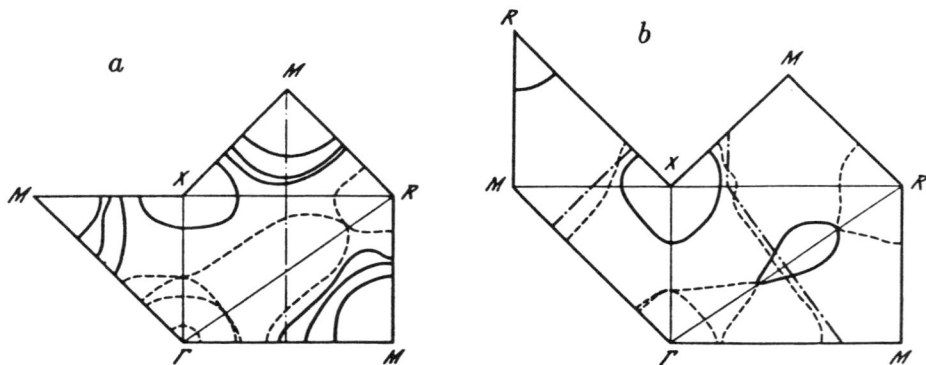

Figure 19   Fermi surfaces for Pt_3Fe (a) and FeRh (b) in the paramagnetic phase. Solid lines: electron part of the surface; Dashed lines: Hole part of the surface; dash-dot lines: boundaries of the Brilouin zone for the antiferromagnetic phases with $\mathbf{k} = [1/2\ 1/2\ 0]$ (for Pt_3Fe) and $\mathbf{k} = [1/2\ 1/2\ 1/2]$ (for FeRh).

matrix-element approximation showed an increase in $\chi^0(q)$ near the points X and M, i. e., that this function is indeed singular at $q = Q_1$ and $q = Q_2$. These results were confirmed by recent experiments on the inelastic scattering of neutrons in paramagnetic Pt_3Fe [39].

3.2. IRON-RHODIUM ALLOYS

It is known from the band theory of magnetism that whether or not the magnetically ordered state is advantageous is determined by the condition $I(q)\chi^0(q, \omega) > 1$, where $I(q)$ is the exchange-correlation parameter (the Stoner integral) and $\chi^0(q, \omega)$ is the polarization operator for the susceptibility, which can be calculated in the self-consistent field approximation. The case $q_1 = 0$ corresponds to a phase transition into the ferromagnetic state at $T = 0$, and since $\chi^0(q_1 = 0, \omega = 0) = N(\epsilon_F)$, the condition for existence of the ferromagnetic phase can be written as

$$I(\epsilon_F)N(\epsilon_F) > 1,$$

which is the Stoner criterion. An analogous inequality,

$$I(\epsilon_F)\chi^0(q_2, 0) > 1,$$

can be obtained for the phase transition into the antiferromagnetic state.

Thus the electronic band structure of FeRh should be characterized by a high density of states in the Fermi level and a peak in the polarization operator $\chi^0(q, 0)$ at some $q_2 = Q$.

The strong-coupling method (with numerous simplifications and assumptions) was used to calculate the electronic structure of the FeRh alloy in Khan [30]. Koenig's [31] calculation was carried out in a self-consistent local approximation using the density functional formalism in the LMTO method. Both calculations were carried out for all three states; however, they contain no information on the topology of the Fermi surface.

Our calculation of the band structure and density of states using 286 points in the 1/48 irreducible part of the Brillouin zone yielded results quite similar to those in [31]. It was found that $N(\epsilon_F) = 124$ states/(Ryd · elem. cell), $I(0) = 0.042$ Ryd, and $I(0)N(\epsilon_F) = 5$, i. e., the ferromagnetic state is preferable to the paramagnetic state at $T = 0$. Figure 18b shows the curve for the density of states in FeRh.

Figure 19b shows the cross sections through the Fermi surface for an iron-rhodium alloy of equiatomic composition in the paramagnetic phase. The Fermi surface consists of two closed hole-type pockets centered on the point Γ, a closed hole pocket centered on the point R, an electron sphere at the point Y, and a deformed electron-type ellipsoid with major axis in the ΓR direction. There is also an open conical hole-type surface with the axis lying along the ΓX direction. This is the largest part of the FeRh Fermi surface by volume.

It has been shown experimentally that the elementary cell of the simple cubic alloy changes into a fcc cell with half the period during the phase transition into the antiferromagnetic phase. The dash-dot lines in Fig. 19b indicate the boundaries of the new Brillouin zone. The figure indicates that the boundaries of the reconstructed Brillouin zone coincide practically everywhere with the flat sections of the Fermi surface, i. e., we have a situation similar to that occurring in the three-dimensional Peierls instability. All portions of this sheet on the Fermi surface move together under translation by the vector $\mathbf{Q} = \pi/a[111]$, which means that the polarization operator has a singularity at this value of $\mathbf{Q}$. An antiferromagnetic gap appears at the boundary of this new zone, and most of the Fermi surface vanishes, which is what leads to the experimentally observed large decrease in the density of states in the Fermi level in the antiferromagnetic phase. The SDW with $\mathbf{Q}$ has nodes on the rhodium atoms, which explains the absence of magnetic noments on the latter in the antiferromagnetic phase.

Thus, the electron band structure of iron–rhodium alloys is conducive to both ferromagnetic and antiferromagnetic ordering. The occurrence of one of these types of magnetic ordering will depend on temperature, pressure, and composition. For example, pressure will lead to enlargement of the zones and, consequently, a decrease in the density of states $N(\epsilon_F)$. This means that the Stoner criterion becomes weaker, and the FM free energy gain upon the establishment of ferromagnetic order is reduced. On the other hand, small deviations from stoichiometric composition will suppress the phase transition into the SDW state, since distortion of the open hole sheet on the Fermi surface will lessen the extent to which the sheet coincides with the boundaries of the Brillouin zone, i. e., the gain in free energy upon formation of the SDW will decrease. However, in alloys with $c_{Fe} \geq 51.5$ at%, the antiferromagnetic phase is apparently in a latent state, and can actually appear under strong hydrostatic pressure.

And so, our experimental study of iron–platinum and iron–rhodium alloys showed that all of the physical properties are extremely strong functions of the composition and thermodynamic properties such as temperature, pressure, and magnetic field. It was fotnd that high pressures make the antiferromagnetic state stable and strong magnetic fields make the ferromagnetic state stable over the entire tempearature range where large-scale magnetic order exists.

Under certain conditions, it was also possible to induce magnetic structures which do not occur under normal conditions.

This instability of the magnetic states in iron–platinum and iron–rhodium alloys could be explained within the framework of the band theory: calculations indicated that it was due to singularities in the topology of the Fermi surface and in the density of states on the Fermi surface in these alloys.

REFERENCES

1. Pál, L., G. Zimmer, J. C. Pichoche, et al., "The magnetic field dependence of the antiferromagnetic–ferromagnetic transition temperature in FeRh," Acta Phys. Hung., vol. 32, pp. 135–140, 1972.
2. Zavadskii, E. A., and I. G. Fakidov, "Magnetic properties of alloys in strong magnetic fields," Fiz. Tverd. Tela, vol. 9, pp. 139–144, 1967.
3. McKinnin, J. B., D. Melville, and E. W. Lee, "The antiferromagnetic–ferromagnetic transition in iron–rhodium alloys," J. Phys. C: Solid State Phys., vol. 1, supplement, pp. S40–S58, 1970.
4. Ponomarev, B. K., "A study of the antiferro–ferromagnetic transition in the alloy FeRh in pulsed magnetic fields of up to 300 kOe" Zh. Éksp. Teor. Fiz., vol. 63, pp. 199–204, 1972.
5. Vinokurova, L. I., V. G. Veselago, V. Yu. Ivanov, et al., "A study of monocrystals of ordered iron–platinum alloys. I. Magnetic properties in strong magnetic fields," Fiz. Metallov i Metallovedenie, vol. 45, pp. 287–293; "II. Anisotropy of the magnetic properties," Fiz. Metallov i Metallovedenie, vol. 45, pp. 869–873, 1978.
6. Kondorskii, E. I., Zonnaya teoriya magnetizma, Ch. I, Ch. II (The Band Theory of Magnetism, Parts I and II), Izd. Moskovsk. Gos. Univ., Moscow, 1976, 1977.
7. Coles, B. R., "Transition from local moment to itinerant magnetism as a function of composition in alloys," Physica B, vol. 91, pp. 167–169, 1977.
8. Sumiyama, K., and C. M. Graham, "Magneto-volume effect in ordered Pt_3Fe alloy," Solid State Comm., vol. 19, pp. 241–243, 1976.
9. Wohlfarth, E. P., "Contributions to the invar problem. II," Phys. Lett. A, vol. 28, pp. 569–570, 1969.
10. Edwards, D. M., and E. P. Wohlfarth, "Magnetic isotherms in the band model of ferromagnetism," Proc. Roy.Soc. London A, vol. 303, pp. 127–137, 1968.
11. Besnuns, M. L., Y. Coffehren, and G. Hunschy, "Magnetic properties of Ni–Cr alloys," Phys. Status Solidi B, vol. 42, pp. 597–607, 1972.
12. Stoner, E. S., "Collective electron ferromagnetism," Proc. Roy. Soc. London A, vol. 165, pp. 372–414, 1938.
13. Dawes, D. G., and B. R. Coles, "A calorimetric investigation of the emerging ferromagnetism in AuFe alloys," J. Phys. F: Metal Phys., vol. 9, pp. L215–L220, 1979.
14. Wohlfarth, E. P., "Collective electron treatment of transition from antiferromagnetism to ferromagnetism in metals," Phys. Lett., vol. 4, pp. 83–84, 1963.
15. Wohlfarth, E. P., and P. Rhodes, "Collective electron metamagnetism," Phil. Mag., vol. 7, p. 1817, 1962.

16. Wohlfarth, E. P., "High magnetic field effects in some metallic magnetic materials," J. Magn. and Magn. Materials, vol. 20, pp. 77–83, 1980.

17. Schinkel, C. J., R. Hartog, and F. H. Hochstenbach, "On the magnetic and electrical properties of nearly equiatomic ordered FeRh alloys," J. Phys. F: Metal Physics, vol. 4, pp. 1412–1422, 1974.

18. Sh. Sh. Abel'skii, "Thermoelectromotive force as a function of temperature for antiferromagnetic metals near the Néel temperature," Fiz. Tverd. Tela, vol. 15, pp. 1414–1416, 1973.

19. Dik, E. G., and Sh. Sh. Abel'skii, "Anomalies in the thermoelectromotive force in ferromagnetic metals near the Curie point," Fiz. Metallov i Metallovedenie, vol. 37, pp. 1305–1308, 1974.

20. Sze, N. H., and G. T. Meaden, "A new effect at first-order phase transitions discovered by a new thermoanalytical technique," Phys. Lett. A, vol. 37, pp. 393–394, 1971.

21. Khvostantsev, L. G., L. F. Vereshchagin, and A. P. Novikov, "Device of toroid type for high pressure generation," High Temp. High Pressure, vol. 9, pp. 637–639, 1977.

22. E. G. Ponyatovskii, A. R. Kustar, and G. T. Dubovka, "The possible existence of a special triple point in the P-T diagram of the alloy FeRh," Kristallografiya, vol. 12, pp. 79–83, 1967.

23. Wayne, R. C., "Pressure dependence of the magnetic transitions in FeRh alloys," Phys. Rev., vol. 170, pp. 523–527, 1968.

24. Zakharov, A. I., A. M. Kadomtsev, R. Z. Levitin, et al., "Magnetic and magnetic ordering characteristics of metamagnetic iron–rhodium alloys," Zh. Éksp. Teor. Fiz., vol. 46, pp. 2003–2010, 1964.

25. Dubovka, G. T., "Effect of pressure on magnetic transitions in iron–rhodium alloys," Zh. Éksp. Teor. Fiz., vol. 65, pp. 2282–2288, 1973.

26. Grazhdankina, N. P., I. F. Mirsaev, and G. G. Taluts, "On the theory of first-order magnetic phase transitions under hydrostatic pressure. II. Conditions for stability of the phases in transitions between disordered and ordered states," Fiz. Metallov i Metallovedenie, vol. 52, pp. 36–43, 1981.

27. Leger, J. M., C. Susse, and B. Vodar, "Point triple dans les diagrammes de phases P-T de deus alliages á base de Fer et ade Rhodium," Compt. Rend. Acad. Sci., vol. 265, pp. 892–895, 1967.

28. Petrov, A. N., "Phase transitions in a phenomenological s-d exchange model," Fiz. Metallov i Metallovedenie, vol. 53, pp. 581–584, 1982.

29. Hargitai, Cs., "On the aligned magnetic moment of the Rh atoms in the FeRh alloys," Phys. Lett., vol. 17, pp. 178–179, 1965.

30. Khan, M. A., "Band theory of non-magnetic and magnetic iron–rhodium alloy," J. Phys. F: Metal Phys., vol. 9, pp. 457–472, 1979.

31. Koenig, C., "Self-consistent band structure of paramagnetic, ferromagnetic, and antiferromagnetic ordered FeRh," J. Phys. F: Metal Phys., vol. 12, pp. 1123–1137, 1982.

32. Itskevich, E. S., "A high-pressure chamber for work at low temperatures," Pribory i Tekhnika Éksperimenta, no. 6, pp. 161–164, 1966.

33. Pál, L., T. Tarnóczi, P. Szabó, et al., "Investigation of afm–fm transition in iron–rhodium alloys," In: Proc. Int. Conf. Magnetism (Nottingham, 1964), Physical Society, Nottingham, pp. 158–161, 1966.

34. Vinokurova, L. I., A. V. Vlasov, and M. Pardavi-Horváth, "Pressure effects on magnetic phase transitions in FeRh and FeRhIr alloys," Phys. Status Solidi B, vol. 78, pp. 353–357, 1976.

35. Wohlfarth, E. P., "The invar problem," IEEE Trans. Magn., vol. 11, pp. 1638–1644, 1975.

36. Kulikov, N. I., "An approximate method for calculating the band structure of a transition metal," Izv. Vuzov Chern. Metallurgiya, vol. 7, pp. 128–132, 1975.

37. Kozlov, A. N., and L. A. Maksimov, "The metal–dielectric phase transition. Bivalent crystals," Zh. Éksp. Teor. Fiz., vol. 48, pp. 1184–1193, 1965.

38. Kourov, N. I., Yu. N. Tsiovkin, S. M. Podgornykh, et al., "Heat capacity of atom-by-atom ordered $(Pd_x Pt_{1-x})_3 Fe$ below 20 K," Fiz. Metallov i Metallovedenie, vol. 8, pp. 81–86, 1982.

39. Kohgi, M., and Y. Ishikawa, "Paramagnetic scattering in neutrons from a metallic antiferromagnet $FePt_3$," J. Phys. Soc. Japan, vol. 49, pp. 994–999, 1980.

KINETIC PROPERTIES OF IRON–RHODIUM ALLOYS IN THE TRANSITION REGION FROM ANTIFERROMAGNETISM TO FERROMAGNETISM

L. I. Vinokurova and V. Yu. Ivanov

Abstract The electrical resistance, magnetoresistance, and Hall effect were studied in ordered monocrystalline iron–platinum alloys with compositions in the transition region between antiferromagnetism and ferromagnetism ($24.8 \leq c_{Fe} \leq 36$ at%) at magnetic fields of up to 150 kOe. It was shown that the kinetic properties as a function of T, H, and c_{Fe} depend strongly on the nature of the magnetic structures, and show anomalies during magnetic phase transitions. The laws governing the variations in the kinetic effects (especially the Hall effect) are indicative of the changes in the band structure of the alloys as a function of composition and temperature, as well as magnetic field strength (at strong magnetic fields).

Magnetic and neutron diffraction studies have led to a determination of the magnetic T–c phase diagram for ordered Pt–Fe alloys with compositions in the transition region between antiferromagnetism and ferromagnetism (see the preceding articles in this volume). Complex magnetic structures were observed; these structures depended not only on alloy composition and temperature but also on magnetic field and pressure (at strong magnetic fields and high pressure).

The large variety in the magnetic structures of these alloys is due to the formation of induced magnetic moments on the platinum atoms under certain conditions due to either the strong internal exchange fields created by the "additional" iron atoms (above 25%) or strong external magnetic fields.

It is well known that the kinetic properties of transition-metal alloys can provide additional information on both the current carrier scattering mechanism and the extent to which the magnetic moment carriers are localized. We therefore carried out a detailed study of the electrical resistivity and the galvanometric and magnetoelastic properties of the alloys. The results of this research have been discussed in greater detail by Vinokurova and Ivanov [1–3].

Measurement Methods. The electrical resistivity, magnetoresistance, and Hall effect were measured photometrically in a constant field using an R248 potentiometer with sensitivity $2 \cdot 10^{-8}$ V; the data were generally recorded continuously on a two-coordinate PDS-021 chart recorder. The measurements were carried out on the same monocrystalline samples used to study the magnetic properties (the samples had mean dimensions of $0.7 \times 0.7 \times 7$ mm), as well as on several polycrystalline samples.

Electrical contacts consisting of a copper or platinum wire 0.05 mm in diameter were welded to the samples by the electric spark method [4]. The symmetry of the Hall contacts was checked using the R348 potentiometer. The difference in potential between the contacts for a sample current of 100 mA was no larger than $5\,\mu$V at room temperature. The mounted samples were glued (using BF-2 glue) to a Getinaks backing attached to a rotating assembly which allowed the sample to be rotated around the horizontal axis [5]. To compensate for the induction due to lack of stability in the magnetic field, some fraction of the EMF induced in a coil consisting of a single loop of wire mounted perpendicular to the magnetic field adjacent to the sample was subtracted from the signal being studied.

To eliminate unwanted effects, the Hall voltage measurement was carried out four times for the four parallel and antiparallel field and current directions at each fixed magnetic field and temperature value.

Temperatures were determined using Chromel–Alumel (at temperatures of 300–900 K) and copper–Constantan or copper–0.15 at% Fe copper alloy (at temperatures of 4.2–900 K) thermocouples.

The absolute values of the electrical resistivity were measured to within approximately 9%. The errors in the magnetoresistance measurements were generally no greater than 0.1% for point-by-point measurements or 5% when recorded on the chart recorder. The measurement error in the Hall voltage was no greater than 1%, while that in the specific Hall voltage was no greater than 5%. The possible errors in the absolute determination of the Hall were much larger (10–20%), because they were determined graphically.

The static magnetic fields (of up to 150 kOe) were obtained at the "Solenoid" facility (Institute for General Physics, USSR Academy of Sciences).

1. ELECTRICAL RESISTIVITY

1.1. LOW-TEMPERATURE ELECTRICAL RESISTIVITY AS A FUNCTION OF TEMPERATURE AND COMPOSITION

Figure 1 shows the specific electrical resistivity ρ as a function of temperature for mono- and polycrystalline alloys of various composition. Figure 1a implies that in "ideal" antiferromagnetic samples close to the stoichiometric composition

Pt_3Fe only a single anomaly is observed in the function $\rho(T)$: a change in slope at $T = 170\,K$ (the Néel temperature of these alloys). Similar results were obtained in [6].

When the alloy composition deviates from stoichiometric, discontinuities are observed in the $\rho(T)$ curves at the temperatures $T = T_1$ which correspond to the point at which the type of large-scale magnetic order changes (see the first paper in this volume). Approximately 1–2 K of hysteresis is observed at T_1 when the samples are heated and then cooled; this is confirmation of the presence of a first-order phase transition at $T = T_1$.

A "pseudo-Kondo" anomaly occurs for temperatures $T < T_1$ in alloys with $c_{Fe} \geq 30.5\,at\%$: the electrical resistivity decreases with increasing temperature (Fig. 1b). Studies of the electrical resistivity in strong magnetic fields enabled us to observe a very interesting phenomenon: sufficiently strong fields can suppress the anomalies in the electrical resistivity as a function of temperature. The normal variation in $\rho(T)$ returns at fields greater than H_{cr} (H_{cr} is the critical magnetic field corresponding to the change in alloy magnetic structure (see the second paper in this volume)) (Fig. 1b, Curves 2′ and 3′). This anomalous behavior of $\rho(T)$ is only observed over a very small range of concentrations: alloys with $c_{Fe} \geq 33$ and $\leq 29\,at\%$ have the usual dependence of electrical resistivity on temperature with a positive temperature coefficient of resistivity

It should be noted that this anomalous variation in electrical resistivity with a negative temperature coefficient of resistance is observed in precisely the range of compositions where magnetic and neutron diffraction experiments show significant inhomogeneity in the magnetic structure. The highest residual electrical resistivity ρ_0 (measured both at zero magnetic field and for $H = 115\,kOe$) and electrical resistivity at 290 K (see Fig. 4 in [1]).

A similar anomalous variation in electrical resistivity with temperature has also frequently been observed in amorphous systems [7] and in crystalline alloys with severe disruptions in their structural or magnetic order, such as (for example) $Cu_{55}Ni_{45}$ [8]. The fact that the anomaly is suppressed by the magnetic field indicates that the anomaly is magnetic in nature for Pt–Fe alloys with iron concentrations in the range 30.5–32.2 at%, rather than being due to structural disorder [9]. Maxima are frequently observed in the specific resistivity near critical concentrations corresponding to the appearance of homogeneous magnetically ordered structure, as in Cr–Fe [10] and Cu–Ni alloys [11] (for example).

The change in the magnetic structure of iron–platinum alloys resulting from the addition of small amounts of iron to Pt_3Fe results in low-temperature deviations from the Mathiessen rule (DMR) that can be described by the following relation:

$$\Delta\rho(c, T) = [\rho(c, T)_{alloy} - \rho(c, 0)_{alloy}] - [\rho(c, T)_{matrix} - \rho(c, 0)_{matrix}]. \quad (1)$$

(where the resistivity of the alloy with $c_{Fe} = 24.8\,at\%$ was adopted for the matrix resistivity).

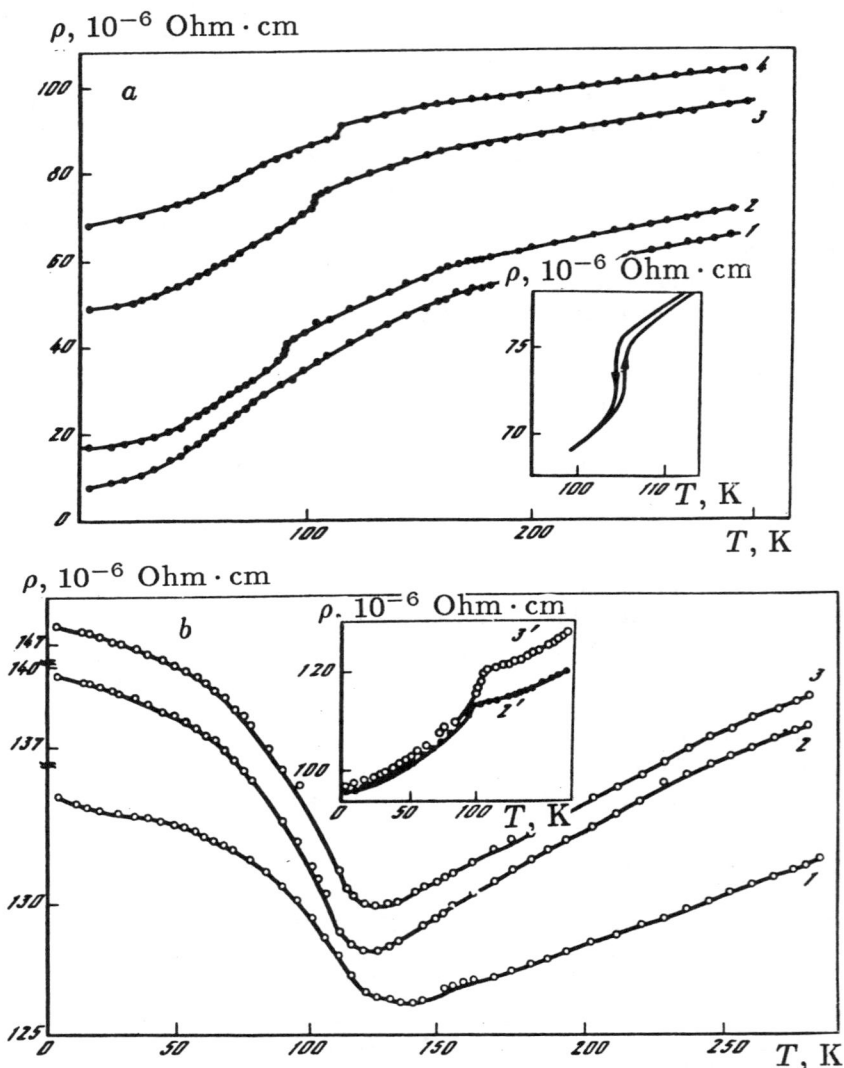

Figure 1   Specific resistivity of iron–platinum alloys as a function of temperature.
(a) Monocrystals with $c_{Fe} < 30\,at\%$ ((1) 24.8, (2) 25.2, (3) 27, (4) 28 at% Fe; inset: $\rho(T)$ in the
vicinity of T_1 for the 27 at% alloy on an enlarged scale); (b) monocrystals with $c_{Fe} > 30\,at\%$
((1) 30.5, (2, 2') 32.16, (3, 3') 30.65 at%; inset: $\rho(T)$ at $H = 115\,kOe$); (c) polycrystals with
29 (1), 33 (2), 34 (3), and 36 at% Fe (4).

The temperature dependence of the DMR implies that the DMR will be most prominent in two regions: at low temperatures and for $T_1 < T < T_2$ (see Fig. 6 in [1]). The DMR increases in size with increasing alloy iron content. The alloys with 30.5–32.2 at% Fe also have large DMRs.

The alloys with 25.2, 27.0, and 28.0 at% Fe are observed to have the characteristic maximum in $\Delta\rho$ of spin glasses and cluster glasses. The presence of this maximum is confirmation of our conclusion that "cluster-glass"-type phases exist in these alloys at low temperatures.

1.2. CURRENT-CARRIER SCATTERING MECHANISMS AT LOW TEMPERATURES

Various methods of analysis have shown that the temperature dependence of the electrical resistivity at low temperatures can best be approximated by an expression of the form

$$\rho_1(T) \equiv \rho(T) - \rho_0 = A_i T^{\alpha_i}. \tag{2}$$

The exponents α_i and, in some cases, the coefficients A_i, were determined by constructing the function $\log \rho_1(T) = f(\log T)$.

The following sequence of values was observed with increasing temperature for α in alloys near the stoichiometric composition Pt_3Fe: $\alpha \approx 2$ for $4.2 \lesssim T \lesssim 10\,K$ (i. e., $T < 0.1\Theta_D$, where Θ_D is the Debye temperature); $\alpha \approx 3.3$–3.6 for $8 \lesssim T \lesssim 22\,K$; $\alpha \approx 2.75$ for $20 \lesssim T \lesssim 40\,K$; $\alpha \approx 1.7$ for $40 \lesssim T \lesssim 80\,K$, and $\alpha \approx 1$ as the temperature approaches T_N. Note that the anomalies become quite sharp in the vicinity of the Néel point.

This pattern of variation in the exponent with temperature is evidently due to competition between different scattering mechanisms. The increase in α to values of order 3.3–3.6 (which is observed only in the alloys of nearly stoichiometric composition) is an indication of the importance of the contribution from scattering on spin waves (which, as was shown in [12, 13] is of the form $\rho \sim T^5$ for antiferromagnets) to the electrical resistivity. The smaller exponent obtained experimentally is apparently due to interference contributions. Thus, the Pt_3Fe alloy can be discussed within the framework of the localized magnetic moment model in accordance with the neutron diffraction results of Ishikawa, et al. [14]. No minimum like that observed in $(Pd_xPt_{1-x})_3Fe$ or $Pt_3Mn_xFe_{1-x}$ for compositions in the transition region between antiferromagnetism and ferromagnetism [15] was observed at temperatures above 2 K. This minimum, which was not suppressed by strong magnetic fields and was evidently due to interference effects in interactions between electrons and in the elastic scattering of conduction electrons on static inhomogeneities, was, however, observed in the partially ordered alloy with 28 at%.

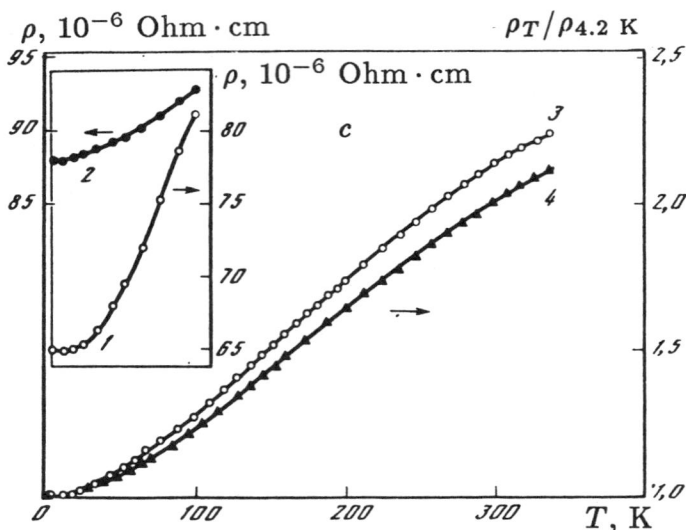

Figure 1   (continued).

The resistivity increases quadratically with temperature for $20 < T < 80\,K$ in the alloys with $c_{Fe} = 25.2–28\,at\%$, and the coefficient A is practically independent of composition and is equal to $(2.2 \pm 0.2) \cdot 10^{-9}\,Ohm \cdot cm/K^2$. Several scattering mechanisms are observed to have quadratic electrical resistivity functions: electron–electron collisions [16, 17], scattering on oscillating impurity ions [18–19a], non-quasi-momentum-conserving scattering on spin waves [20], and scattering on fluctuations in the collectivized d electron spin density [21].

However, in the case where the electron–electron collisions and scattering on spin waves dominate, the coefficient of proportionality should be equal to 10^{-11}–$10^{-12}\,Ohm \cdot cm/K^2$ [22], or two orders of magnitude smaller than the values observed in Pt–Fe alloys. Also, for spin-wave scattering (especially scattering on oscillating impurity spins), the coefficient A should depend on alloy composition.

On the other hand, large values of A were predicted for band antiferromagnets by Ueda [21], and experimentally observed in antiferromagnetic materials such as α–Mn $(A = 110 \cdot 10^{-9})$ [23]; $NiS_{2-x}Se_x$ $(A = (12–33) \cdot 10^{-9})$ [24]; CrB_2 $(A = 1.5 \cdot 10^{-9})$ [25] (in all cases, the values of A are given in $Ohm \cdot cm/K$), which indicates that the contribution from scattering on fluctuations in spin density is substantial in Pt–Fe alloys with the compositions in question. The fact that we are dealing with band magnetism here is even more apparent in the alloys with $c_{Fe} \geq 30.5\,at\%$, where the exponent α is equal to 1.6–1.7 and the coefficient A is a factor of three larger in the alloys with $30.5 \leq c_{Fe} \leq 32.2\,at\%$ for $H = 115\,kOe >$

H_{cr} (where the characteristic dependence of electrical resistivity on temperature for metals returns) and in the alloys with $c_{Fe} \geq 33$ at%, for zero magnetic field and temperatures ranging from 20 to ~ 100 K. These data are presented below, along with the parameters obtained from the function $\rho = \rho_0 + AT^\alpha$:

c_{Fe}, at%	25.2	27.0	28.0	30.65	32.2	33.0	34.0	36.0
α	2.04	2.04	2.02	1.64	1.69	1.7	1.8	1.6
A, 10^{-9} (Ohm · cm)/K$^\alpha$	2.3	2.3	2.1	6.7	6.7	–	–	–
ΔT, K	20–80	20–70	20–60	20–100	20–100	30–100	20–90	25–90

The error in the determination of α was 0.05 for the alloys with $c_{Fe} = 25.2$–32.2 at% and 0.1 for the alloys with $c_{Fe} = 33$–36 at%.

In all cases, α begins to decrease with increasing temperature above 100 K, while it increases with increasing temperature for $T < 20$ K.

This transition from a quadratic temperature dependence at low temperatures to $\alpha = 5/3$ (and 1) with increasing temperature is predicted in Ueda and Moriya's [26] theory of band magnets in the vicinity of the ferromagnetism boundary. It should be noted that the Curie point anomalies are ill-defined in the alloys with $c_{Fe} \geq 33$ at%, and α becomes less than unity at $T > T_C$, also in accordance with this theory. Similar characteristics are also observed in $ZrZn_2$ [27] and Pd–Ni alloys [28], for example.

1.3. HIGH-TEMPERATURE ELECTRICAL RESISTIVITY

Figure 2 shows the electrical resistivity of monocrystalline iron–platinum alloys and pure polycrystalline platinum measured as a function of temperature up to ~ 900 K, i. e., temperatures much greater than the Debye temperature Θ_D of these materials (the values of Θ_D for alloys with $c_{Fe} = 25$, 28, and 32 at% is 258 [29], 252, and 243 K, respectively–the latter values were determined from the temperature dependence of the heat capacity; $\Theta_D = 225$–240 K for platinum [30]).

For $T > \Theta_D$, the quantity $d\rho/dT$ should be determined by the phonon part of the electrical resistivity, and should be independent of temperature; however, as Fig. 2 indicates, $\rho(T)$ is nonlinear even for $T \ll \Theta_D$. The deviations from linearity in $\rho(T)$ are quite large, and cannot be explained by anharmonicity. The derivative $d\rho/dT$ is also observed to vary substantially (by approximately a factor of two) with alloy composition (Table 1); this can hardly be completely due to changes in the alloy phonon spectra, since both the parameters of the crystal lattice [31] and the Debye temperature of the alloys change only slightly with composition. The electronic structures of the alloys evidently undergo significant deformation with changes in alloy composition.

We can attempt to explain these characteristics of the electrical resistivity in iron–platinum alloys using the theory of electrical resistivity developed by Mott [32]. Mott suggested that the more mobile s electrons might be scattered in collisions with impurities, phonons, and electrons in unoccupied d-band states (s–d transitions).

The heat capacity measurements indicated that the coefficient of the contribution to the heat conductivity linear in temperature increases by a factor of 3.3 with increasing iron concentration from 28 to 32 at% ($\gamma = 2.7$ and 9.0 mJ $\cdot$ mole$^{-1} \cdot$ K^{-2}, respectively). It can be assumed that the density of states in the Fermi level also increases with increasing iron concentration. The properties of iron–platinum alloys can then be explained within the framework of the Mott model resulting from an increase in the extent to which electrons in the d band scatter on magnetic and structural inhomogeneities, which leads to an increase in the residual resistivity and a decrease in $d\rho/dT$ with increasing iron concentration.

According to the Mott model, the high-temperature resistivity in the paramagnetic phase should be proportional to $T(1 - AT^2)$ (taking the phonon contribution into account). Analysis showed that the temperature dependence of the electrical resistivity in Pt–Fe alloys for $T > \Theta_D$ can to first approximation be described by the Mott equation, with the coefficient $A = 1.8 \cdot 10^{-7}$ K^{-2}.

However, the low value of the temperature coefficient of the resistivity determined by constructing $(\rho(T) - \rho_0)/(\rho(\Theta) - \rho_0) = f(T/\Theta)$, where $\Theta = T_2$ was adopted for the Pt–Fe alloys and $\Theta = 170$ K $= T_2$ was adopted for the Pt$_3$Fe alloy (see Fig. 10 in [1]) indicates that in contrast with the paramagnetic metals Pt and Pd and the similar metal Ni, the additional high-temperature resistivity in the Pt–Fe alloys is due to the scattering of conduction electrons on disordered "magnetic atom spins" and on phonons.

On the other hand, some of the characteristics of the electrical resistivity in

Table 1. Temperature Coefficient of Resistance, $\partial\rho/\partial T$, 10^{-8} Ohm $\cdot$ cm/K for Pt–Fe Alloys at High Temperatures

c_{Fe}, at%	ΔT, K			
	300–400	400–600	600–800	800–950
24.8	8.4	7.6	6.0	—
24.9	9.2	8.0	6.4	—
27.0	7.2	6.2	5.0	4.3
30.65	3.2	2.8	2.0	1.7
0	4.0	3.4	3.6	3.2

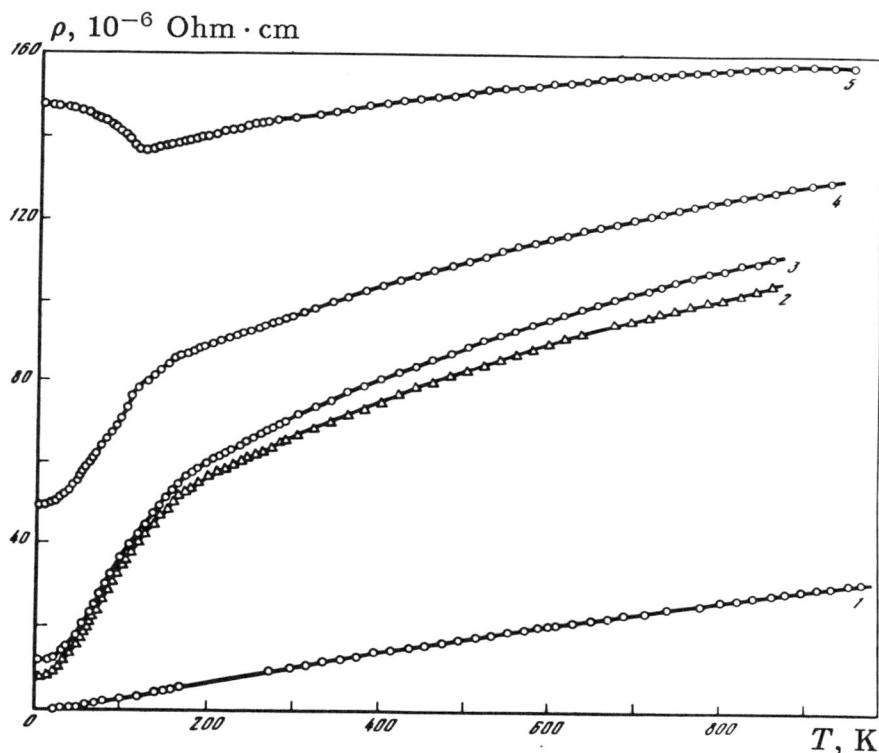

Figure 2   Specific electrical resistivity of monocrystalline iron–platinum alloys and polycrystalline platinum as a function of temperature measured at high temperatures.

iron–platinum alloys can be qualitatively explained within the framework of a localized electron model based on the hybrid interaction between the localized and collectivized electrons discussed in Atoyan, Barabanov, and Maksimov [33]. This interaction may lead to a change in the configuration of the unfilled localized electron shells as the localized electrons undergo direct transitions into the conduction band. This process leads to the appearance of a hybrid contribution to the electrical resistivity in addition to the usual impurity and phonon contributions, which may lead to an increase to values of order $100\,\mu$Ohm $\cdot$ cm for ρ_0 in the strong-hybridization region, deviations from the Mathiessen rule, logarithmic deviations of $\rho(T)$ from linearity at high temperatures, or even a minimum in the curve $\rho(T)$.

The decrease in the temperature coefficient of resistivity with increasing alloy iron content may also be due to the "saturation effect" noted by Allen [34]. When the electron mean free path becomes comparable in order of magnitude to the interatomic separation, the mean free path stops decreasing at the same rate that the scattering cross section cross section increases, and "saturates" at some minimum mean value. The electron–electron, electron–phonon, and electron–impurity collisions are no longer independent, and cannot be described using the Boltzmann equations.

Generalization of the experimental results in [35] indicates that $d\rho/dT$ decreases with increasing resistivity, and becomes negative for $\rho \geq 150\,\mu\text{Ohm} \cdot \text{cm}$. This precise value of the electrical resistivity occurs in the alloys with $c_{\text{Fe}} = 30.5$–32.2 at%. However, as was shown above, the negative temperature coefficient of the resistivity in these alloys, is magnetic in origin.

The experimental data presented above led us to the conclusion that the electrical resistivity of the antiferromagnetic alloy Pt_3Fe (which has magnetic moments only on the iron atoms) as a function of temperature is similar to that observed in alloys with localized magnetic moments, and alloys with compositions in the transition region are observed to have the characteristic behavior of band magnetic materials.

2. MAGNETORESISTANCE

Magnetoresistance—an even galvanomagnetic effect—is defined as the relative change in electrical resistivity in an external magnetic field:

$$\Delta\rho/\rho_0 = \frac{\rho(H, T) - \rho(0, T)}{\rho(0, T)}, \tag{3}$$

where $\rho(H, T)$ and $\rho(0, T)$ are the resistivity at temperature T in an external magnetic field $H \neq 0$ and $H = 0$, respectively.

There are at least two main causes (in addition to the "ordinary" magnetoresistance due to the cyclotron orbital motion of the conduction electrons in the magnetic field as a result of the Lorentz force) for this change in resistivity with magnetic field. First of all, the appearance of large-scale magnetic order significantly affects the conduction electron energy spectrum; this leads to a change in the number of current carriers and their effective masses. Secondly, the magnetic field affects the current-carrier scattering mechanisms (i. e., the current-carrier relaxation time).

Since the magnetoresistance is a reflection of the connection between the magnetic structure and kinetic properties of magnetically-ordered alloys, we studied the magnetoresistance in iron–platinum alloys as a function of temperature and magnetic field strength in magnetic fields of up to 150 kOe on the temperature interval from 4.2 to 290 K, including the phase transition region.

2.1. MAGNETORESISTANCE FOR ALLOYS OF STOICHIOMETRIC COMPOSITION Pt_3Fe ($c_{Fe} = 24.8\text{–}24.9$ at%)

Figure 3 shows the transverse ($\mathbf{I} \perp \mathbf{H}$, where $\mathbf{I}$ is the sample current) magnetoresistance as a function of magnetic field strength at 4.2 K for monocrystalline alloys containing less than 30 at% Fe. The low-temperature magnetoresistance is positive for alloys with the stoichiometric composition Pt_3Fe, and $(\Delta\rho/\rho_0)_\perp > (\Delta\rho/\rho_0)_\parallel$. Increasing temperature leads to a decrease in both the longitudinal ($\mathbf{I}\|\mathbf{H}$) and transverse magnetoresistance. At $H = 100$ kOe, for example, $(\Delta\rho/\rho_0)_\perp$ is 10.4, 1.5, 0.7, and 0.25% for $T = 4.2, 55, 77$, and 91 K, respectively, for the alloy with $c_{Fe} = 24.8$ at%.

In order to correctly isolate the contribution to the magnetoresistance resulting from the antiferromagnetic ordering of the crystal, we must subtract out the contribution from the "ordinary" magnetoresistance. The latter is well known to obey Kohler's rule, which can be written as

$$\Delta\rho/\rho_0 = F(H/\rho_0), \tag{4}$$

where F is a universal function depending on the relative orientations of the field and current in the sample, but not the temperature of the sample or the presence of impurities in the sample.

We subtracted out the "Kohler contribution" to the magnetoresistance in much the same way as described in [23] for α-Mn: namely, we assumed that the difference in magnetoresistance between two samples of nearly stoichiometric composition ($c_{Fe} = 24.8$ and 24.9 at%) having different ratios of resistivity at room temperature to that at 4.2 K (8.75 and 6.15, respectively) can be completely described by the "ordinary part" of the magnetoresistance. In the process, we found that the the difference between the magnetoresistances increased proportional to $H^{3/2}$, rather than quadratically. Similar functions have been observed in the "Kohler construction" for several metals and alloys, including platinum [37].

The magnetoresistance in this case can be represented by the following function:

$$\Delta\rho/\rho_0 = B(H/\rho_0)^{3/2} + (\Delta\rho/\rho_0)_{AFM}. \tag{5}$$

A value of $1.19 \cdot 10^{-12}$ (Ohm $\cdot$ cm/kOe)$^{3/2}$ was obtained for B.

Figure 4 shows the "antiferromagnetic" contribution to the magnetoresistance obtained by subtracting the "Kohler part" of the magnetoresistance from the total magnetoresistance. The "antiferromagnetic contribution" at 4.2 K for the alloys

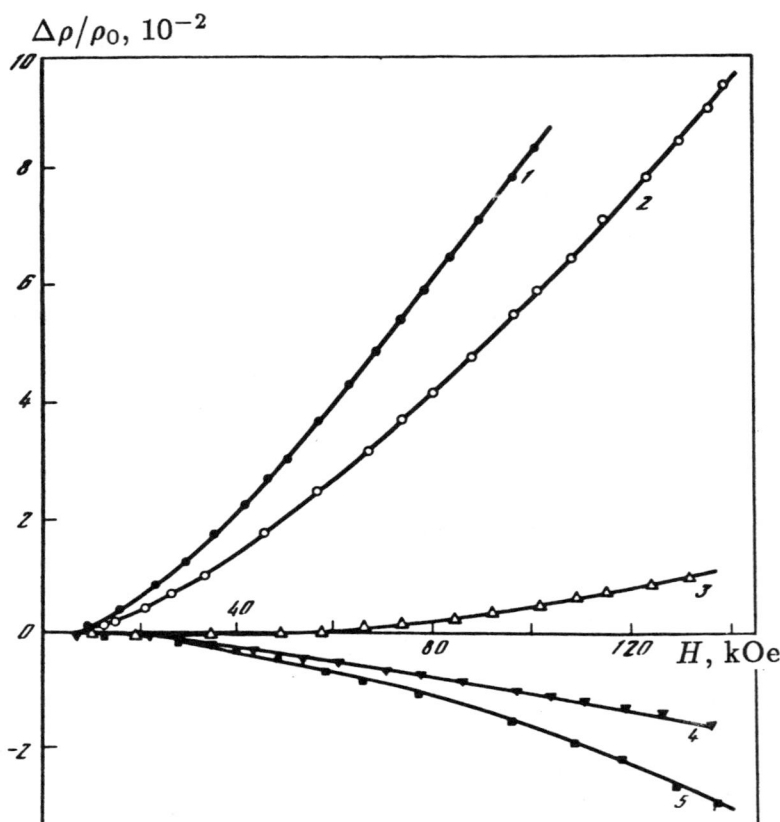

Figure 3   Transverse magnetoresistance for iron–platinum alloys of various composition as a function of magnetic field strength. c_{Fe}, at%: (1) 24.8, (2) 24.9, (3) 25.2, (4) 27, (5) 28.

with $c_{Fe} = 24.8$ and 24.9 at% is positive, and amounts to some 50% of the total magnetoresistance. However, in contrast to the case for α–Mn, the dependence on H is nonlinear, and cannot be written in terms of a simple power law; this may be because the "Kohler part" was not subtracted out quite correctly.

The magnetoresistance is lower at 77 K, so that the "antiferromagnetic contribution" obtained becomes particularly sensitive to the size of the "Kohler part." The inaccuracy in the determination of the latter was evidently due to the different signs for the "antiferromagnetic contribution" in the alloy with 24.9 at% Fe (positive contribution) and 24.8 at% Fe (negative).

$(\Delta\rho/\rho_0)_{\text{AFM}}, 10^{-2}$

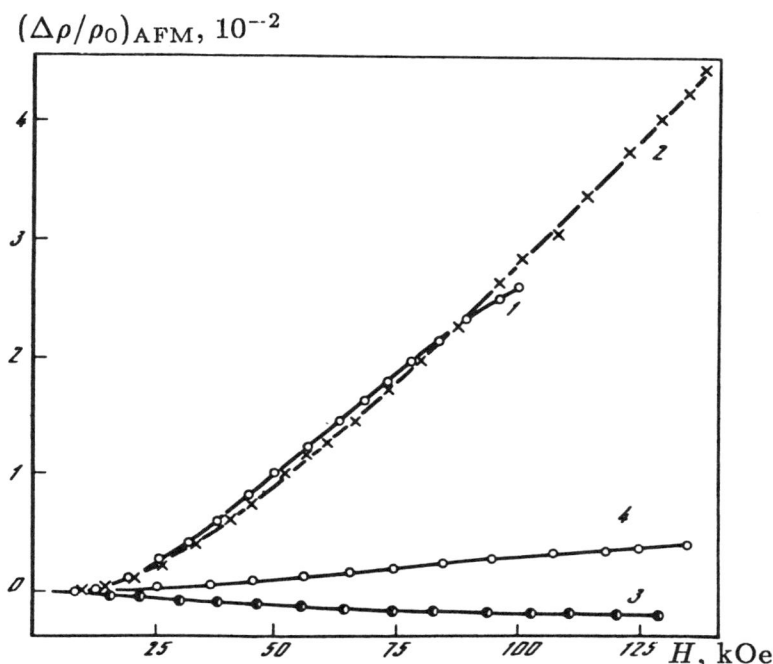

Figure 4 "Antiferromagnetic contribution" to the magnetoresistance of iron–platinum alloys of stoichiometric composition as a function of magnetic field strength. c_{Fe}, at%: (1, 3) 24.8, (2, 4) 24.9; T, K: (1, 2) 4.2, (3, 4) 77.

The data presented above imply either that the antiferromagnetic contribution changes sign or that there is an additional negative contribution (larger than the "Kohler" and the "antiferromagnetic" contributions) which increases in size with temperature at higher temperatures. The latter idea seems more likely to us. One possible source for this additional contribution is the negative contribution to the magnetoresistance from interference between the non-magnetic ("Kohler") and magnetic scattering of conduction electrons [38].

To date, only a limited volume of work on the theory of magnetoresistance in antiferromagnets has been carried out. This theoretical work does not provide the basis for an adequate quantitative description of the behavior of the magnetoresistance in iron–platinum alloys of stoichiometric composition. A calculation by Yamada and Takada [38] using the localized magnetic moment model indicated that the effects of the magnetic field on the electrical resistivity in a two-sublattice antiferromagnet are a result of a change in the spin density wave spectrum. Usami [39] has discussed the scattering of current carriers (s electrons) on d electron spin fluctuations in weak band antiferromagnets. Both theories predict a linear decrease in electrical resistivity with magnetic field for small magnetic fields ($\mu_B H \ll k_B T$), which is not experimentally observed in the Pt–Fe alloys. However, in the localized magnetic moment theory, the negative effect is replaced by a positive effect at strong magnetic fields ($k_B T \ll \mu_B H \ll \mu_B H_{c2}$, where H_{c2} is the "collapse field strength" for the antiferromagnetic sublattices. According to [39], $\rho(T)$ should decrease with increasing field strength for $T > T_N$, and there should be a tendency for saturation in strong fields, as is observed in Pt$_3$Fe.

It should be noted that the temperature dependence of the magnetoresistance in MnAu$_3$, a ferromagnet with localized magnetic moments, is similar to that of $\Delta\rho/\rho_0$ in Pt$_3$Fe: $\Delta\rho/\rho_0 > 0$ for $T < T_N$; the magnetoresistance changes sign near T_N, and reaches a minimum at $T = T_N$ [40].

2.2. MAGNETORESISTANCE IN ALLOYS OF NONSTOICHIOMETRIC COMPOSITION ($25 < c_{Fe} < 30.5$ at%) IN FIELDS LESS THAN H_{cr}

The functions $\rho(H)$ show different behavior when the iron concentration exceeds 25 at% (see Fig. 3): $\Delta\rho/\rho_0$ decreases in absolute value, changes sign, and then becomes negative at 27 at%. Even at helium temperatures, $(\Delta\rho/\rho_0)_{\parallel}$ and $(\Delta\rho/\rho_0)_{\perp}$ are negative for fields of up to 50 kOe (where the contribution due to the "ordinary" magnetoresistance is small) in the samples with $c_{Fe} = 25.2$ at%. On the other hand, the magnetoresistance is negative in the alloys containing 27 and 28 at% Fe independent of the field strength.

We attempted to obtain an approximate estimate of the size of this negative magnetoresistance by subtracting the "Kohler term" from the total magnetoresistance, as we did for the stoichiometric-composition alloys. The curves obtained for $(\Delta\rho/\rho_0)_{\text{add'l}} = (\Delta\rho/\rho_0)_{\text{total}} - (\Delta\rho/\rho_0)_{\text{Kohler}}$ at 4.2 K are shown in Fig. 5. The magnetoresistance for the alloy with $c_{Fe} = 25.2$ at% also becomes negative after subtracting the "Kohler term." As the iron concentration increases, $(\Delta\rho/\rho_0)_{\text{add'l}}$ as a function of magnetization approaches a quadratic, finally becoming one at $c_{Fe} = 28$ at%. Note that $\Delta\rho/\rho_0$ also differs only slightly from being quadratic with respect to the field (and is also nearly quadratic with respect to magnetization) at other temperatures as well.

The magnetoresistance is also theoretically predicted to have similar properties ($\Delta\rho/\rho_0 < 0$ and $\Delta\rho/\rho_0 \sim J^2$) within the framework of the s–d model in dilute alloys with magnetic impurities [41–43], and these properties have been observed in many ferromagnetic and antiferromagnetic alloys with deviations from order, such as, for example, disordered ferromagnetic Pd–Fe alloys ($c_{Fe} = 1.0$–2.7 at%) [44] and dilute alloys with "magnetic" impurity atoms [41]. This behavior is due to the scattering of conduction electrons on impurity spins. Thus, the main contribution to the magnetoresistance in Pt–Fe alloys which deviate from stoichiometric composition comes from the ordering of magnetic inhomogeneities in the external field.

2.3. MAGNETORESISTANCE IN STRONG MAGNETIC FIELDS NEAR PHASE TRANSITIONS

We shall now discuss the magnetoresistance as a function of magnetic field in the temperature region where "discontinuities" in the magnetic field are observed at $H = H_{cr}$.

Figure 6a shows the electrical resistivity of a 25.2 at% Fe alloy in a longitudinal field as a function of magnetic field strength for various temperatures in the vicinity of T_1. The curves for transverse field are similar.

The size of the discontinuity in electrical resistivity, $\Delta\rho'$, is nearly independent of temperature over the narrow range of temperatures studied, and amounts to some 10% of $\rho(0, T)$. The function $\rho(H)$ falls off monotonically with increasing field above T_1.

With increasing iron concentration, the discontinuities in $\rho(H)$ at $H = H_{cr}$ decrease, and are observed to be anisotropic. In the samples with $c_{Fe} = 27$ at%, $\Delta\rho'/\rho_0$ is 4.6 and 4.1% in transverse fields with $\mathbf{H}\|[110]$ and $\mathbf{H}\|[100]$, respectively, and 1.5% in a longitudinal field directed along the [001] axis. In contrast to the case for the alloy with $c_{Fe} = 25.2$ at%, $\Delta\rho'$ does depend on temperature, but only very slightly.

An additional characteristic is observed in the alloys with 28 at% Fe: the sign of the discontinuity in $\rho(H)$ changes from positive to negative with decreasing temperature. Figure 6b shows the functions $\rho(H)$ measured in a longitudinal magnetic field. It should be noted that the change in sign occurs at different temperatures in longitudinal and transverse fields: at $T \approx 93$ K in a longitudinal field parallel to the [110] axis and at 83 and 65 K in transverse fields $\mathbf{H}\|[110]$ and [111], respectively. However, $\Delta\rho'$ remains positive down to 60 K (the lowest possible temperature for observing $\Delta\rho'$ for $H = H_{cr}$ up to 150 kOe) for $\mathbf{H}\|[001]$, and there is no trend towards a change in sign.

A similar pattern of changes in the sign of the discontinuity in electrical resistivity at $H = H_{cr}$ has been observed in the magnetoresistance of dysprosium with helical magnetic structure region [45].

Figure 6c shows the functions $\rho(H)$ in longitudinal fields for the 30.65 at% Fe alloy. For $T > 27$ K, the monotonic decrease in $\rho(H)$ in this alloy is replaced by saturation at $H = 30–40$ kOe; this is evidently due to the fact that the clusters become oriented by the external magnetic field at this temperature. The discontinuity in $\rho(H)$ at $H = H_{cr}$ is negative for all temperatures in the alloys with $c_{Fe} \geq 30.5$ at% Fe, while the size of the discontinuity decreases with increasing

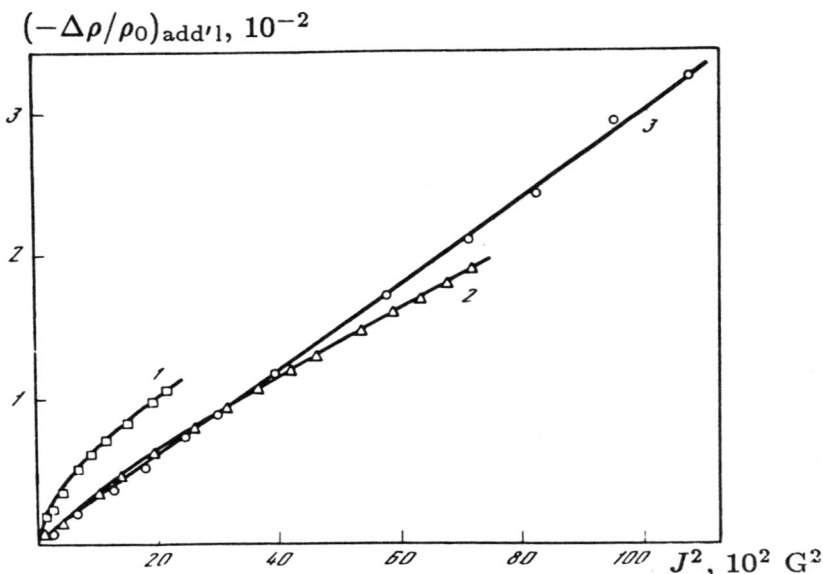

Figure 5   Additional contribution to the magnetoresistance in alloys of nonstoichiometric composition as a function of the square of the magnetization. c_{Fe}, at%: (1) 25.2, (2) 27.0, (3) 28.0.

temperature (for $30\,K < T < T_1$). The size of the discontinuity in the electrical resistivity in these alloys also depends on the direction of magnetization.

"Remanence effects" were observed as the field decreased out of the strong-field regime: ρ at $H = 0$ turned out to be smaller once the field was turned off than it was before application of the field, which confirms that there are "frozen clusters" in the alloys at low temperatures.

The existence of discontinuities in the electrical resistivity at $H = H_{cr}$ is not surprising, since it is a first-order phase transition. Large discontinuities in the electrical resistivity were observed to accompany the discontinuities in magnetization for the metamagnetic phase transitions in Fe–Rh alloys were observed to be accompanied by large discontinuities in the electrical resistance ($\Delta\rho < 0$) [46] and $Co(S_xSe_{1-x})_2$ ($\Delta\rho$ positive, and equal to approximately $60\,\mu Ohm \cdot cm$ independent of concentration or temperature) [47].

Current theories (both the localized [38] and the collective magnetic moment [39] models) predict the existence of discontinuities in the $\rho(H)$ curves during first-order phase transitions (which is precisely the case for the phase transitions at $H = H_{c1}$, where H_{c1} is the magnetic sublattice inversion field); the localized mag-

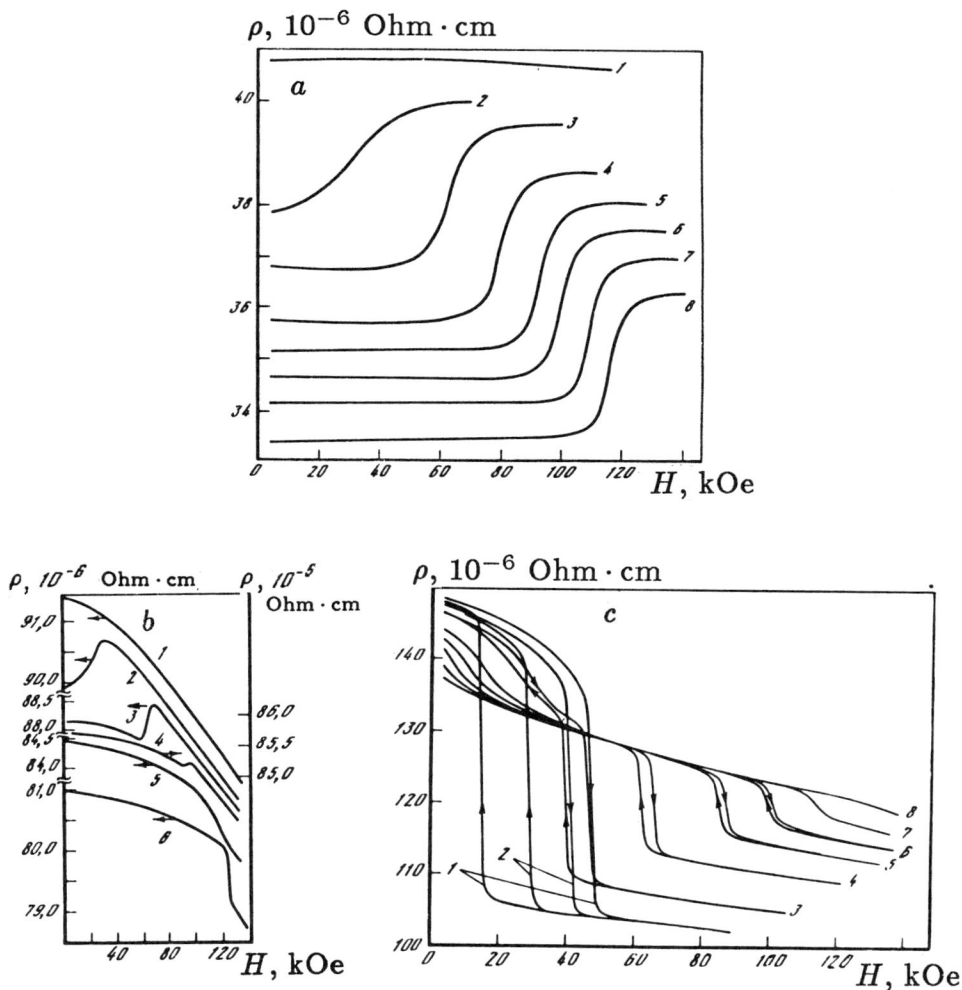

Figure 6 Electrical resistivity isotherms for iron–platinum alloys in a longitudinal field. (a) $c_{Fe} = 25.2$ at% and $H\|I\|[110]$ (T, K: (1) 93.75, (2) 89.25, (3) 87, (4) 84.75, (5) 83; (6) 81.75, (7) 80, (8) 78); (b) $c_{Fe} = 28$ at% and $H\|I\|[110]$ (T, K: (1) 117.5, (2) 116, (3) 109.5, (4) 90.5, (5) 94, (6) 78); (c) $c_{Fe} = 30.65$ at% and $H\|I\|[100]$ (T, K: (1) 4.2, (2) 27, (3) 52.25, (4) 74.5, (5) 88.5, (6) 95, (7) 100.5, (8) 106).

netic moment model predicts a negative sign for the discontinuity $\Delta\rho'$, the band model yields a positive sign.

The transitions at $H = H_{cr}$ are similar to those at $H = H_{c1}$ in the alloys with $c_{Fe} = 25.2$–28 at%. The phase transitions observed in iron–rhodium alloys with iron concentrations greater than 30.5 at% are similar to those discussed by Moriya and Usami [48], and should also be accompanied by similar anomalies in $\rho(H)$ [39].

However, the change in sign of the discontinuity in electrical resistivity at $H = H_{cr}$ with changing alloy composition, and the change in sign with temperature in the alloy with $c_{Fe} = 28$ at% cannot be clearly explained within the framework of the theories discussed above.

The behavior of the magnetoresistance in the critical-field region can be qualitatively explained if magnetostrictive crystal distortions are taken into account using the phenomenological expression suggested in [45]. When a magnetic field is applied along the axis α, the magnetoresistance measured in direction i at constant pressure p can be written in the following form:

$$\left(\frac{\partial \rho_i}{\partial H_\alpha}\right)_p = \sum_j \left(\frac{\partial \rho_i}{\partial L_j}\right)_{H_\alpha} \left(\frac{\partial L_j}{\partial H_\alpha}\right)_p + \left(\frac{\partial \rho_i}{\partial H_\alpha}\right)_{L_j}, \tag{6}$$

where the L_j are lattice parameters, the first term characterizes the magnetostrictive contribution to the magnetoresistance, and the second term characterizes the electrical resistivity due to changes in the spin configuration and electronic structure.

If the $(\partial \rho_i / \partial L_j)_H$ are monotonic, we then have

$$\Delta \rho_i = \sum_j \left(\frac{\partial \rho_i}{\partial L_j}\right)_H \Delta L_j + (\Delta \rho_i)_{L_j}. \tag{7}$$

Our results concerning the effect of pressure on the electrical resistivity of iron–rhodium alloys imply that ρ decreases under hydrostatic compression, i. e., that $(\partial \rho / \partial L)_H > 0$. The sign of the first term in equation (7) is therefore determined by the sign of the lattice deformation at $H = H_{cr}$. Our measurements of the magnetostrictive deformations of crystals at $H = H_{cr}$ imply that $\Delta L_j > 0$ in in the alloy with $c_{Fe} = 25.2$ at%, while ΔL_j can be of either sign in the alloys with high iron concentrations.

As for the sign of the second term in (7), nearly-ferromagnetic alloys ($c_{Fe} \geq 30.5$ at%) undergo a phase transition from a magnetically inhomogeneous state (antiferromagnet + clusters) into a coherent structure in which the clusters are no longer free, but "coupled" to the direction of magnetization by not only the external field but also by the strong internal fields. There should be less scattering on magnetic inhomogeneities in this state, i. e., $(\Delta \rho_i)_{L_j} < 0$. This is confirmed by the presence of a substantial decrease in electrical resistivity, even at fields less than the critical field for partial cluster ordering. The second term of (7) dominates over the first term for alloys with these compositions; this is what in fact determines the sign and temperature dependence of the discontinuity in electrical resistivity (since $\Delta \rho_i(T)$ is related to $\Delta J(T)$). The anisotropy in the discontinuities $\Delta \rho$ is evidently also due to the anisotropic nature of the crystal deformation at $H = H_{cr}$.

The sign of the discontinuity $(\Delta \rho_i)_{L_j}$ cannot be clearly predicted in the alloys with 25.2–28 at% Fe. However, the tendency for the magnetic scattering to be lower in the resulting weakly ferromagnetic phase will clearly prevail as the discontinuity at $H = H_{cr}$ increases in size. The contributions from both terms (taking the anisotropy in the crystal deformations into account) are likely to be especially important in the alloy with $c_{Fe} = 28$ at%.

2.4. MAGNETORESISTANCE AS A FUNCTION OF TEMPERATURE

Figure 7 shows the magnetoresistance of the iron–platinum alloy with 27 at% Fe measured as a function of temperature in longitudinal magnetic fields of 60 and 100 kOe. Similar results were also obtained for the alloy with $c_{Fe} = 28$ at%.

Anomalous behavior was observed in the vicinity of the magnetic phase transitions near T_1 and T_2. The large positive effect on the interval $T' \leq T \leq T_1$ is due to the first-order phase transition at $H = H_{cr}$, which is accompanied by discontinuities in the electrical resistivity. The size of the effect $\Delta \rho / \rho_0$ in this region is determined by the ratio of the values of $\Delta \rho / \rho_0$ for $H < H_{cr}$ and $H > H_{cr}$, as well as the size of the discontinuity, $\Delta \rho'$. The width of the region, $T' - T_1$, is determined by the magnetic fields at which the measurements were carried out. This region owes its existence to the fact that the phase transition at $T = T_1$ is shifted to lower temperatures in strong magnetic fields $(H = H_{cr})$, i. e., T' is the temperature of the phase transition for the magnetic field where the measurement was made.

At lower temperatures, all of the variations in $\rho(H)$ occur in the region $H < H_{cr}$. This low-temperature region is characterized by a negative magnetoresistance which decreases in absolute magnitude fairly rapidly with increasing temperature up to $T = T'$. An ill-defined maximum in $|\Delta \rho / \rho_0|$ is observed for $T = 7$–10 K. This maximum may be regarded as being due to the fact that the clusters freeze out at these temperatures. The second anomaly-free region is for $T_1 < T < T_2$, where the magnetoresistance is much larger in absolute value than for $T < T'$. Similar functions were also observed for the magnetization and susceptibility. This is evidently due to the formation of magnetic structure with a low ferromagnetic moment above T_1.

The transverse magnetoresistance has the same qualitative behavior as the longitudinal resistivity at low temperatures $(T < T_2)$. An analysis of the variation in magnetoresistance for fields in different directions indicated that the transverse

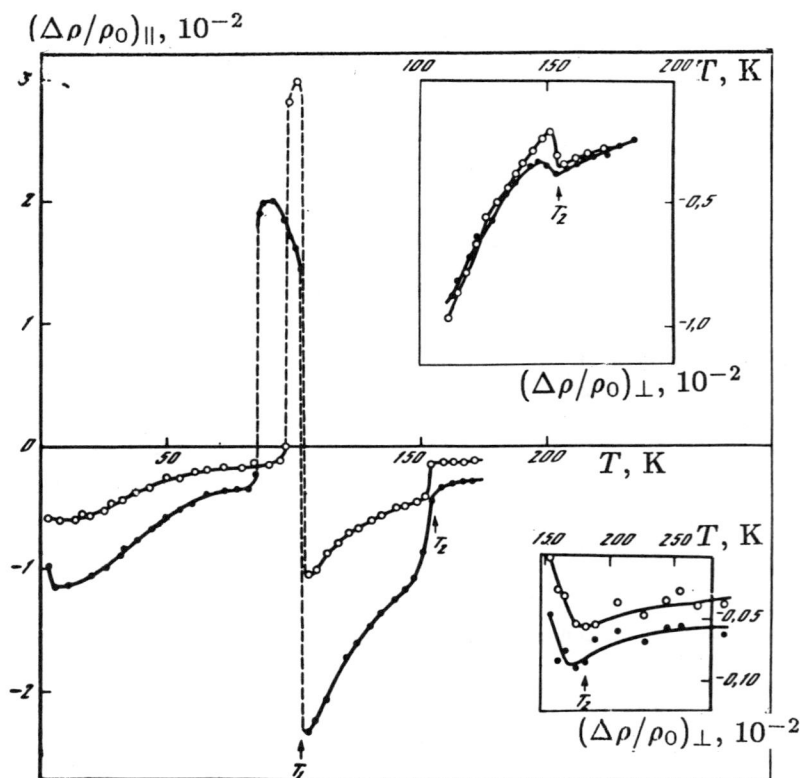

Figure 7 Longitudinal magnetoresistance for a iron–platinum alloy with 27 at% Fe measured as a function of temperature in magnetic fields of 60 and 100 kOe. Insets: Transverse magnetoresistance in the vicinity of T_2 as a function of temperature: (top) 27 at% Fe at $H = 100$ kOe (open circles: **H**∥[110], filled circles: **H**∥[100]); (bottom) 24.8 at% Fe at $H = 60$ (open circles) and 100 kOe (filled circles).

magnetoresistance in alloys with $c_{Fe} = 27$–28 at% reaches a maximum when the magnetic field is directed along a ⟨110⟩-type crystallographic axis. This led us to assume, in accordance with the magnetic measurements, that this is the direction of easy magnetization for alloys of this composition.

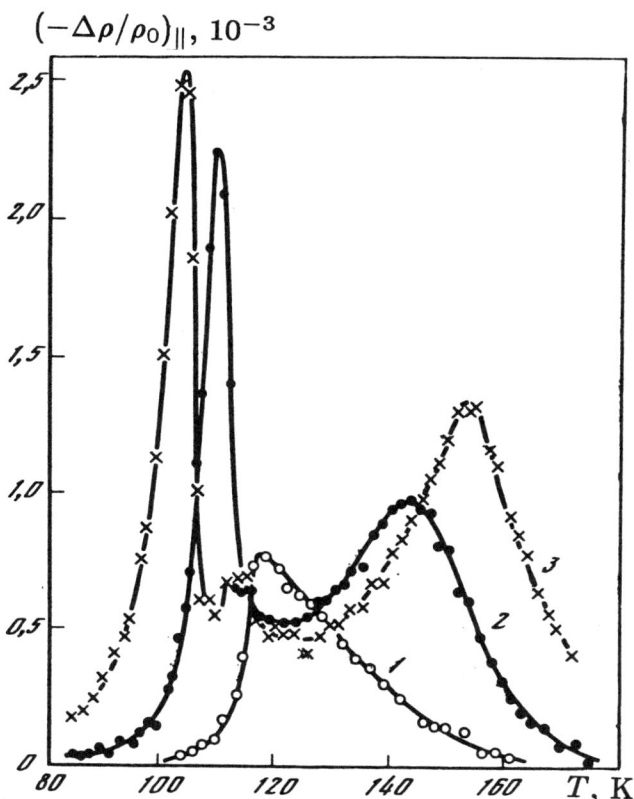

Figure 8   Longitudinal magnetoresistance for iron–platinum alloys in a weak magnetic field ($H = 250\,\mathrm{Oe}$) as a function of temperature. c_{Fe}, at%: (1) 30.5, (2) 30.65, (3) 32.2.

The insets in Fig. 7 show the transverse magnetization along various crystallographic axes for alloys with $c_{\mathrm{Fe}} = 27.0$ and 24.8 at%. Some anomalies are apparent in the vicinity of the Néel point, even though there are no well-defined anomalies in the magnetic properties at high magnetic fields ($H > 10\,\mathrm{kOe}$) at $T = T_N$. No anomalies in $\Delta\rho/\rho_0$ were observed at T_N in [49]. This may be due to the fact that the effect is small in this alloy, and the small fields ($H \leq 16\,\mathrm{kOe}$) used in this study.

A sharply-defined peak in the magnetoresistance at the Néel point due to fluctuations in the small-scale order near the magnetic ordering point has been observed to exist in rare-earth metals [50] and the alloy $MnAu_3$ [40] (for example). One characteristic of iron–platinum alloys near the Néel temperature is that the critical fluctuations in the short-range order are small, with the result that the anomaly in $\Delta\rho/\rho_0$ at T_N takes the form of an ill-defined peak. The characteristics of the anomaly depend on the mutual orientation of the current and magnetic field, as well as the crystallographic axis along which the field is applied.

This behavior of the magnetoresistance in the vicinity of the Néel temperature can also be qualitatively explained on the basis of equation (6) above. The difference between the values of $(\partial L/\partial H_\alpha)_p$ and $(\partial\rho/\partial H_\alpha)_{L_j}$ for $T < T_N$ and $T > T_N$ leads to a different slope in $(\partial\rho/\partial H_\alpha)_p$, and, thus, to a different value for $\Delta\rho/\rho_0$ above and below the magnetic ordering point. The anisotropy in $(\Delta\rho/\rho_0)_\perp = f(T)$ is evidently due to the anisotropic deformation of the lattice which occurs as the temperature decreases below T_N.

It is interesting to note that we did not observe any splitting of the anomalies at T_2 like that observed in the vicinity of $T' - T_1$. This confirms that there is no noticeable shift in the Néel temperature in magnetic fields of up to at least 100 kOe.

The measurements of the differential susceptibility in alloys with $c_{Fe} \geq 30.5$ at% indicated that the anomalies at the magnetic phase transition points are quite strongly suppressed by strong magnetic fields (see the first paper in this volume). To study this further, we carried out measurements of the magnetoresistance in weak longitudinal magnetic fields $(H = 250\,Oe)$. The results of the measurements are shown in Fig. 8. It is apparent that the anomalies observed in $(\Delta\rho/\rho_0)_{\parallel}$ for weak fields are consistent with the anomalies in the susceptibilities of these alloys.

For example, only a single maximum is observed in $\Delta\rho/\rho_0$ at $T = 117\,K$ for $c_{Fe} = 30.5$ at%, while two maxima are clearly visible in the alloy with 30.65 at% Fe. Just as in the χ_d measurements, three maxima were observed in the alloys with $c_{Fe} = 32.2$ at% at temperatures T_1, T_2, and T_3, corresponding to the following transitions: paramagnetism–ferromagnetism–canted structure–antiferromagnetism + clusters.

The data presented above imply that measuring the weak-field magnetoresistance is just as sensitive a method for identifying magnetic phase transitions as susceptibility measurements in weak constant magnetic fields using a variable current.

2.5. MAGNETORESISTIVE CHARACTERISTICS OF ALLOYS WITH $c_{Fe} \geq 30.5$ at% IN FIELDS $H > H_{cr}$

We shall now discuss some of the magnetoresistive characteristics of alloys with over 30 at% Fe that behave like typical ferromagnets at $H > H_{cr}$.

The functions $\rho(H)$ for alloys with these compositions are well-correlated with the functions $J(H)$ in these alloys. It is well known that $\Delta\rho/\rho_0 \sim J^2$ for ferromagnets in both the induced magnetization region and the true magnetization region. The functions $\Delta\rho/\rho_0 \sim J^2$ are satisfied on various sections of the curves in Fig. 9, and remain the same for $T > T_1$.

If we separate out the part of the effect due to the true magnetization by subtracting the value obtained by extrapolation of the $\Delta\rho/\rho_0 = f(H)$ curve from the region $H > H_{cr}$ to $H = 0$ from the total magnetoresistance $\Delta\rho/\rho_0$, the x-intercept will yield a value of J_s^2 which is, to within the errors, in agreement with the value obtained from the magnetization curves directly (Table 2).

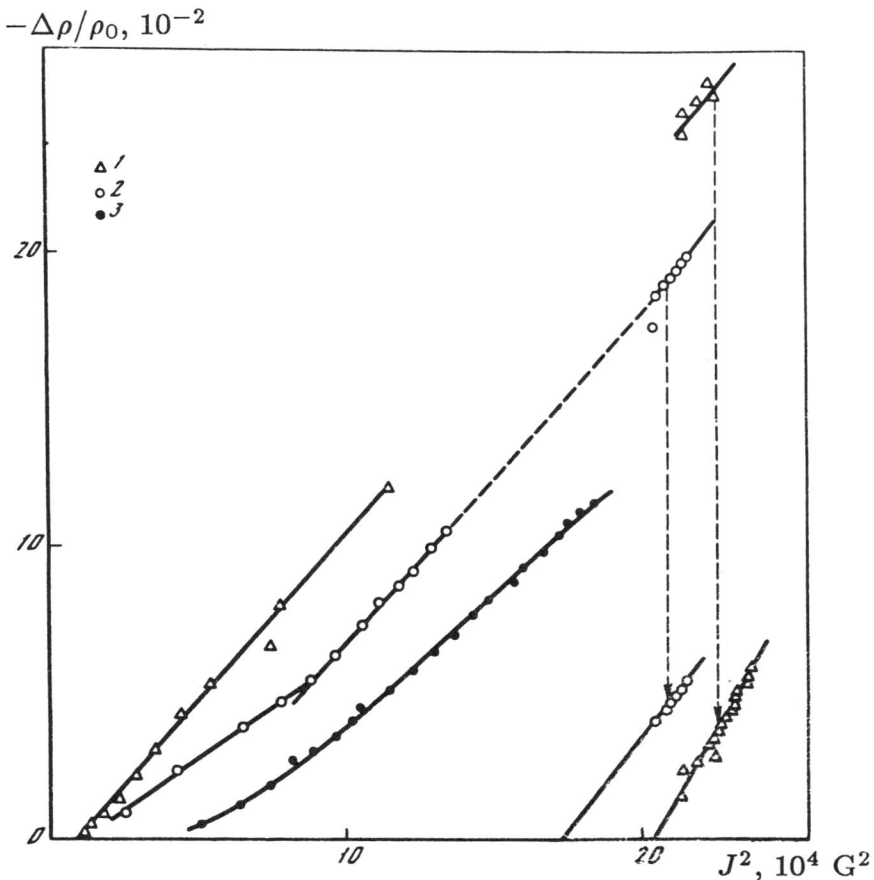

Figure 9   Transverse magnetoresistance as a function of the square of the magnetization for the alloy with $c_{Fe} = 32.2$ at% ($I\|[001]$, $H\|[100]$). T, K: (1) 4.2, (2) 77, (3) 125.

Table 2. Residual Magnetization Values
Determined from Magnetic and Galvanomagnetic Measurements

c_{Fe}, at%	T, K	J_s, G	
		from $J(H)$	from $\Delta\rho/\rho_0$
30.5	4.2	410 ± 5	405 ± 10
	77	380 ± 10	372 ± 10
	125	0	~ 95
32.2	4.2	450 ± 10	450 ± 10
	77	400 ± 10	415 ± 10
	125	250 ± 20	240 ± 10

The alloys with $c_{Fe} \geq 30.5$ at% have noticeable anisotropy in the magnetoresistance for $H > H_{cr}$ in addition to the anisotropy of the discontinuities in the
electrical resistivity. Figures 10a and 10b show the rotation diagrams of the electrical resistivity measured in the (001) and (110) planes for a 90-kOe field.

The electrical resistivity as a function of angle can be described in terms of
the following expression, which takes into account both the terms of second and
fourth order in the cosines of the spontaneous magnetization vector $J_s(\alpha_1, \alpha_2, \alpha_3)$
and electric current density $j(\beta_1, \beta_2, \beta_3)$ [51]:

$$\Delta\rho_s/\rho_0 = \kappa_1 \left(\sum_i \alpha_i^2 \beta_i^2 - \frac{1}{3} \right) + \kappa_2 \sum_{i \neq j} \alpha_i \alpha_j \beta_i \beta_j$$

$$+ \kappa_3 \left(\frac{1}{2} \sum_{i \neq j} \alpha_i^2 \alpha_j^2 - \frac{1}{3} \right) + \kappa_4 \left[\sum_i \alpha_i^4 \beta_i^2 + \frac{1}{3} \sum_{i \neq j} \alpha_i^2 \alpha_j^2 - \frac{1}{3} \right] \quad (8)$$

$$+ \frac{1}{6}\kappa_5 \sum_{i \neq j \neq k} \alpha_i \alpha_j \alpha_k^2 \beta_i \beta_j,$$

where the κ_i are constants and $\Delta\rho_s/\rho_0 \equiv (\rho_{B \to 0} - \rho_0)/\rho_0$ is the value of the
magnetoresistance at saturation, extrapolated to zero induction.

Taking the fact that the magnetoresistance is isotropic in the true magnetization
region into account, we thus see that the electrical resistivity can be written in
the following form as a function of the angles θ and ϕ (where θ and ϕ are the

$\rho,\ 10^{-6}\ \mathrm{Ohm\cdot cm}$

$\rho,\ 10^{-6}\ \mathrm{Ohm\cdot cm}$

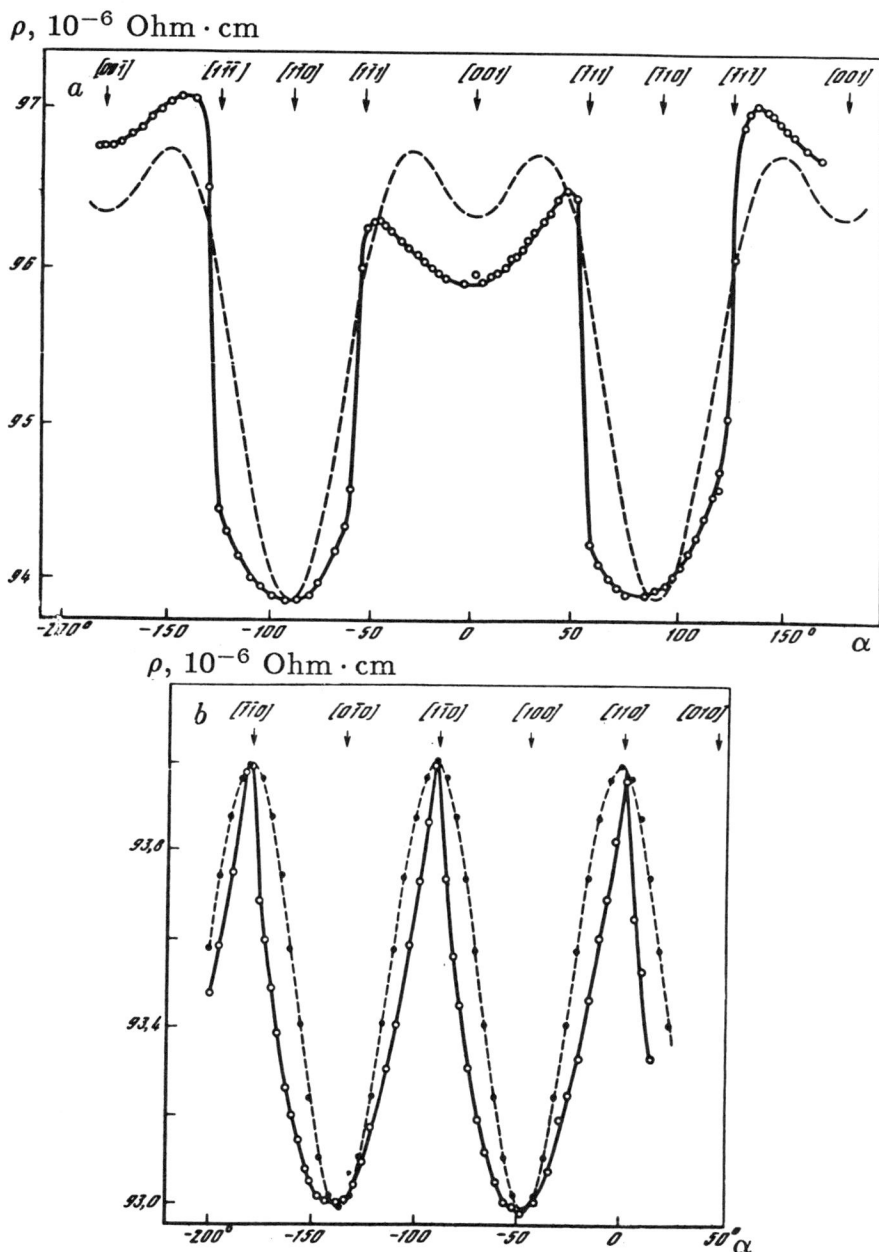

Figure 10   Electrical resistivity measured in a 90-kOe field at 4.2 K as a function of angle for the alloy with $c_{Fe} = 30.5$ at% as the field rotates in the (110) plane (a) and for the alloy with $c_{Fe} = 32.2$ at% when the field rotates in the (001) plane. The dashed curves were calculated using equations (9) and (10).

angles between the vector J_s and the [001] axis in the (110) and (100) planes, respectively):

$$\rho_{H=\text{const}} = A + B\sin^2\theta + C\sin^4\theta, \tag{9}$$

$$\rho_{H=\text{const}} = A' + D\cos 4\phi. \tag{10}$$

Figs. 10a and 10b also show the curves calculated using these equations. The mean extremal values of the electrical resistivity taken from the experimental curves in Fig. 10 and the condition on the point of inflection in the (110) plane $(\partial\rho/\partial\theta = 0$, which implies $B + 2C\sin^2\theta = 0)$ were used to determine the parameters A, B, C, A', and D.

The theoretical curves provide a correct qualitative reproduction of the variation in electrical resistivity with angle in the magnetic field; however, they are smoother functions of angle. For example, they do not reflect the sharp anomalies in the vicinity of the $\langle 111 \rangle$ axes. Similar sharp anomalies were observed near the point where $\mathbf{H} \| \langle 111 \rangle$ in the "rotation curves" for magnetostrictive distortions along various crystallographic axes. These effects may be regarded as being due to a discontinuous change in the orientation of the magnetic moments when the magnetic field passes near $\langle 111 \rangle$-type axes.

And so, the magnetoresistance in alloys with the stoichiometric composition Pt_3Fe has a specific positive contribution associated with antiferromagnetic ordering and the effect of the magnetic field on the spin wave spectrum. The negative contribution to the magnetoresistance (which is proportional to the square of the alloy magnetization and is a result of the ordering of magnetic inhomogeneities by the external field) begins to dominate with increasing iron concentration. In strong magnetic fields, alloys with iron concentrations greater than 30.5 at% have the characteristic properties of ferromagnetic metals and alloys in strong magnetic fields.

In addition to this, the magnetoresistance also shows anomalies in both the first- and second-order phase transitions, i. e., it is a sensitive indicator of changes in alloy magnetic structure.

3. THE HALL EFFECT

Studying the Hall effect—an odd galvanomagnetic effect—is a very fruitful method for studying magnetically ordered metals and alloys in which the ordinary component due to the action of the Lorentz force on the current carriers moving in the magnetic field is accompanied by an anomalous component due to the presence of spontaneous magnetization. The idea that the extraordinary Hall effect is due to the asymmetric scattering of current carriers due to the spin–orbit interaction can be considered firmly established [52].

Since the various scattering mechanisms for the current carriers will in principle make different contributions to the AHE as a result of various types of SOI, analyzing the dependence of the extraordinary Hall effect on temperature and composition can provide valuable information on the dominant scattering mechanisms and the extent to which the magnetic moment carriers in metals and alloys are localized.

Monocrystalline samples of iron–rhodium alloys with 24.8–32.2 at% Fe were used to study the Hall effect at temperatures ranging from 4.2 to 340 K in fields of up to 150 kOe.

3.1. PROCEDURE FOR DETERMINING
THE HALL COEFFICIENTS

The Hall resistivity or specific Hall EMF E_X (the latter term is more frequently used in the Soviet literature) was calculated using the formula $E_X = U_X d/i$, where U_X is the potential difference perpendicular to the external magnetic field H_z and the current i_y, and d is the thickness of the sample in the z direction.

As was found experimentally by Pugh [53, 53a] and Kikoin [54], the specific Hall EMF in ferromagnets is determined by not only the external magnetic field but also by the magnetization, and it can be written in the following form for $T < T_C$ (where T_C is the Curie temperature):

$$E_X = R_0 B + R_s J, \tag{11}$$

where B is the induction in the sample $(B = H + (4\pi - N)J)$; R_0 and R_s are the ordinary and extraordinary Hall coefficients, respectively; H is the external magnetic field; and N is the sample magnetization factor. Thus,

$$E_X = R_0 H + [R_0(4\pi - N) + R_s] J. \tag{12}$$

Assuming $dR_0/dH = dR_s/dH = 0$, we find that

$$dE_X/dH = R_0 + [R_0(4\pi - N) + R_s] dJ/dH. \tag{13}$$

There are two methods of calculating R_0 and R_s from equations (12) and (13):

1. By constructing the functions $E_X = f(H)$. Estimates indicate that $R_0(4\pi - N) \ll R_s$, so that this term can be neglected. In this case,

$$R_s = E_X(H = 0)/J_s, \tag{14}$$

where $E_X(H = 0)$ is determined by extrapolation from the region $H > H_s$, and

$$R_0 = dE_X/dH(H > H_s) - R_s dJ/dH. \tag{15}$$

For $H > H_s$, $J = J_s + \chi_p H$, where J_s is the spomtaneous magnetization at a given temperature and χ_p is the susceptibility in the true magnetization region. Then, $dJ/dH = \chi_p$. However, it has been shown experimentally that the susceptibility

χ_p is not strictly constant (even for fields greater than the critical field, where the function $E_X(H)$ is nearly linear); because of this, average values of χ_p were used to calculate the coefficients. This method was used to determine R_0 and R_s for $H > H_{cr}$ and $T < T_1$.

2. Equation (12) implies that

$$E_X/H = R_0 + [(4\pi - N)R_0 + R_s] J/H. \tag{16}$$

Constructing $E_X/H = f(J/H)$ immediately leads to values for R_0 and R_s. This method of determining the coefficients leads to a smaller error in the determination of R_s, since there is no need to carry out the extrapolation to $H = 0$ or use a mean value for χ_p. Also, in several cases, the functions $E_X = f(J/H)$ turned out to be linear for both $H < H_{cr}$ and $H > H_{cr}$, as well as for temperatures above T_C in the alloys with $c_{Fe} \geq 30.5$ at%, where the curves for $E_X(H)$ are nonlinear. However, the errors in graphically determining R_0 by extrapolation to $J/H = 0$ are larger for this method.

The construction $E_X = f(J/H)$ can be used not just at constant temperatures but also for various temperatures in constant magnetic fields. This method of determining R_0 and R_p (R_p is the extraordinary Hall coefficient in the paramagnetic region) was suggested by Kikoin [55] for determining the Hall coefficients in the paramagnetic temperature region.

3.2. THE HALL EFFECT IN ALLOYS WITH $c_{Fe} \geq 30.5$ at%

The Hall Coefficients. In order to identify the contributions to the Hall EMF from the ordinary and extraordinary Hall effects, we shall now discuss the function E_X at various temperatures as shown for the alloy with $c_{Fe} = 32.2$ at% in Fig. 11. As the figure indicates, the shape of the $E_X(H)$ curves in the vicinity of $T_1 - T_2$ is analogous to that for ferromagnets (Curve 3); however, complete saturation does not occur, even at strong fields ($H > 110\,\mathrm{kOe}$). The function E_X is practically linear for $H > H_{cr}$ and $T < T_1$, while the E_X curves have negative slope (Curves 1 and 2).

It should be noted that hysteresis effects are observed in the function $E_X(H)$ in the vicinity of $H = H_{cr}$.

Figures 12a and 12b show R_0 and R_s for the alloy with 32.2 at% Fe as a function of temperature. The values of the Hall coefficients are given in Table 3. The figures and table indicate that the ordinary Hall coefficient is negative over the entire temperature interval where it was possible to determine the coefficients, while the extraordinary Hall coefficient is positive.

One interesting characteristic of the alloys with $c_{Fe} \geq 30.5$ at% is that the Hall coefficients undergo an abrupt change at $H = H_{cr}$: R_0 decreases by a factor of 2–3 in absolute value (depending on the temperature), while R_s decreases by a factor of 1.5–2. The changes in R_0 and R_s are especially large at low temperatures. These

E_X, 10^{-7} Ohm · cm

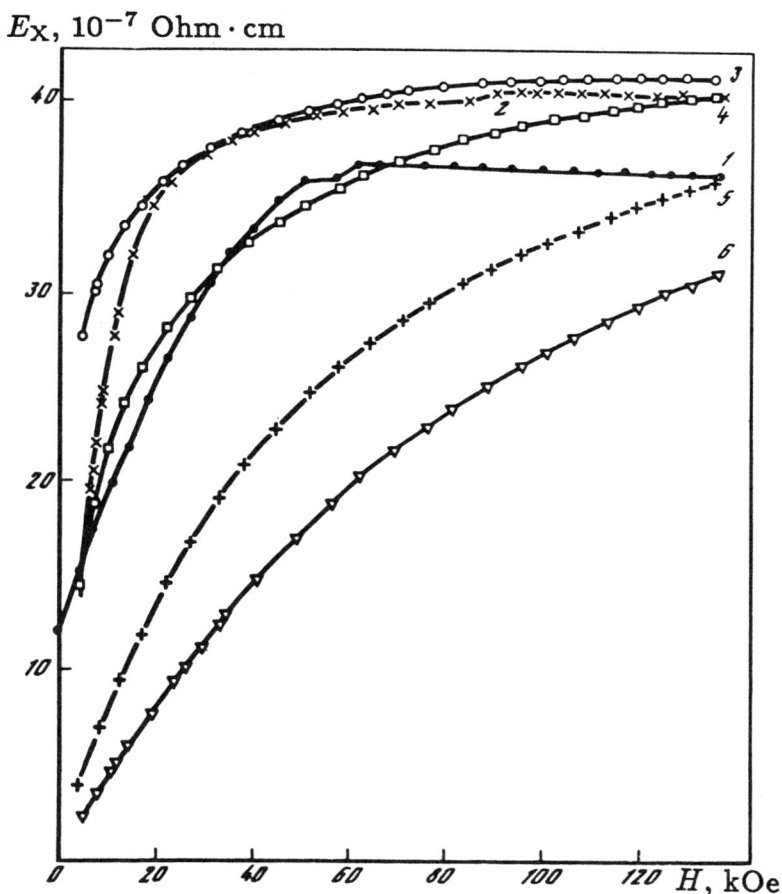

Figure 11 Specific Hall EMF as a function of magnetic field strength for the alloy with 32.2 at% Fe. T, K: (1) 4.2, (2) 77, (3) 100, (4) 183, (5) 247, and (6) 291.

data indicate that both the magnetic and the band structure of the alloys undergo a change at $H = H_{cr}$. We obtained the following temperature-independent values of the coefficients in the paramagnetic region for $H = 11\,\text{kOe} = \text{const}$ by constructing $E_X/H = f(J/H)$: $R_0 = -6.0 \cdot 10^{-12}\,\text{Ohm} \cdot \text{cm/G}$ and $R_s = 13.3 \cdot 10^{-9}\,\text{Ohm} \cdot \text{cm/G}$, values similar to those obtained at stronger fields. Although R_0 also remains constant for $T > T_C$ in strong fields ($H > 40\,\text{kOe}$), R_s continues to increase with temperature, even for $T > T_C$.

$-R_0$, 10^{-12} Ohm $\cdot$ cm/G

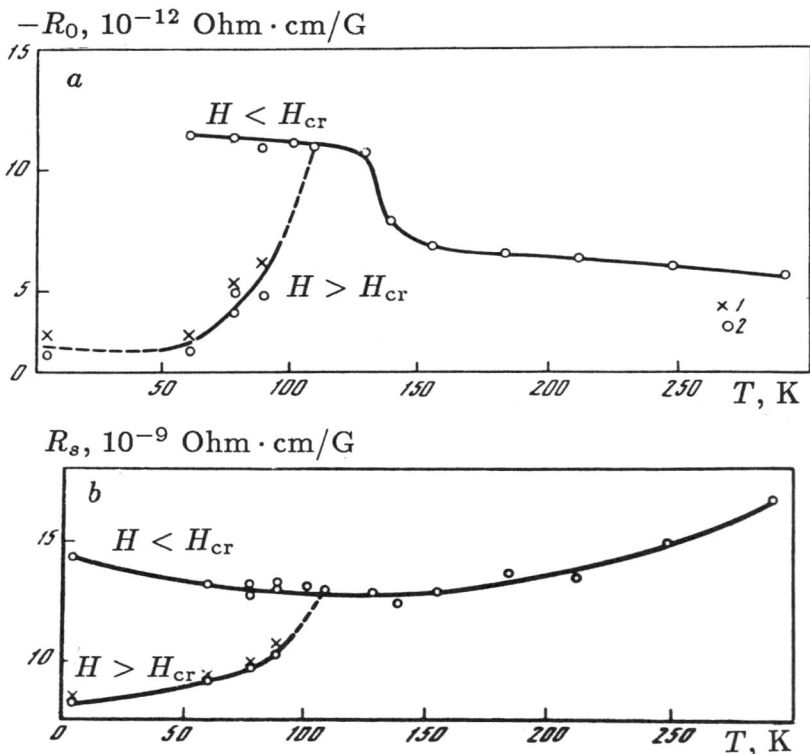

R_s, 10^{-9} Ohm $\cdot$ cm/G

Figure 12   Hall coefficients R_0 (a) and R_s (b) for the alloy with 32.2 at% Fe as a function of temperature ($\mathbf{H}\|[110]$, $\mathbf{I}\|[001]$). Coefficients R_0 and R_s determined using the following functions: (1) $E_X = f(H)$ and (2) $E_X/H = f(J/H)$.

It is also interesting to note that the extraordinary Hall coefficient does not show any anomalies at the transition through the Curie point and that the absolute value of the ordinary coefficient decreases with increasing temperature. Peculiarities are observed in the ordinary Hall coefficient near the magnetic ordering point in many ferromagnetic metals and alloys. This can be explained in terms of the characteristics of the electron system in magnetically ordered metals—specifically, the change in the number of conduction electrons at the onset of spontaneous magnetization [56].

Similar results were also obtained for the alloy with iron concentration 30.5 at%.

We did not carry out a special study of the anisotropy in the Hall effect. However, measurements were made at temperatures of 4.2 and 77 K with the magnetic

Table 3. Hall Coefficients in a Pt–Fe Alloy with 32.2 at% Fe

T, K	Magnetic Field Direction	$H < H_{cr}$		$H > H_{cr}$		Method of Coefficient Determination
		R_0, 10^{-12} Ohm· cm·G^{-1}	R_s, 10^{-9} Ohm· cm·G^{-1}	$-R_0$, 10^{-12} Ohm· cm·G^{-1}	R_s, 10^{-9} Ohm· cm·G^{-1}	
				$T < T_1$		
4.2	[100]	—	—	3.5	7.4	1
		—	10.5	3.0	7.3	2
	[110]	—	—	3.3	8.3	1
		—	14.4	2.5	8.2	2
		2.9	17.6	—	—	3
60	[110]	—	—	3.4	9.3	1
		11.5	13.25	2.7	9.2	2
77	[100]	—	—	4.6	8.6	1
		11.4	11.1	4.4	8.55	2
	[110]	—	—	5.1	9.9	1
		15.0	13.35	4.3	9.8	2
	[1̄10]	—	—	5.5	9.9	1
		13.3	12.8	5.0	9.8	2
88	[110]	—	—	6.35	10.8	1
		11.0	13.1	4.9	10.4	2
				$T > T_1$		
108	[110]			11.0	13.05	2
128	[110]			10.8	12.9	2
				10.3	12.8	3
138	[110]			8.0	12.5	2
154	[110]			7.0	13.0	2
183	[110]			6.7	13.9	2 ($H > 10\,\mathrm{kOe}$)
211	[110]			6.65	13.65	2 ($H > 30\,\mathrm{kOe}$)
247	[110]			6.2	15.2	2 ($H > 40\,\mathrm{kOe}$)
291.5	[110]			5.9	17.0	2 ($H > 70\,\mathrm{kOe}$)

NOTE: Method of coefficient determination: (1) using the function $E_X = f(H)$; (2) using the function $E_X = f(J/H)$; and (3) least squares calculation.

field in various crystallographic orientations. Table 3 also contains the values of the Hall coefficient for these cases. There is no anisotropy in the Hall coefficient to within the experimental errors.

An Analysis of the Scattering Mechanisms Leading to the Anomalous Hall Effect.   The most important part of analyzing the extraordinary Hall effect is to determine what model describes the electrons responsible for the onset of magnetic order. This model determines which of the various scattering mechanisms and types of spin-orbit interaction are responsible for the extraordinary Hall effect. In the band model, for example, the extraordinary Hall effect is due to the internal spin–orbit interaction $(d_{spin}-d_{orbit})$ when magnetized electrons scatter on impurities, lattice defects, electrons, and phonons. In the localized magnetic electron model, on the other hand, the main contribution to the extraordinary Hall effect comes from the scattering of the non-magnetic electrons on magnetic ions due to the mixed spin–orbit interaction (d_{spin}–s_{orbit}) and the internal intra-ion spin–orbit interaction [52].

Current theories indicate that different current-carrier scattering mechanisms lead to slightly different temperature dependences for the extraordinary Hall coefficient. The situation is complicated somewhat in alloys by contributions to the extraordinary Hall effect arising from interference between various scattering mechanisms. In addition, there is a scattering mechanism-independent contribution to the extraordinary Hall effect (which has come to be called the "side jump") in concentrated alloys and metals at high temperatures.

Berger [57] has given a simple physical interpretation of this process. He showed that, in contrast to ordinary "skew" scattering, where there is some angle between the electron trajectories before and after scattering, the side jump involves a shift $\Delta y \approx 10^{-10}$–$10^{-11}$ m in the center of mass of the wave packet, while the trajectories before and after scattering remain parallel.

Iron–rhodium alloys with $c_{Fe} \geq 30.5$ at% have a high specific resistivity: $\sim$ 140 μOhm · cm, which indicates that this contribution to the extraordinary Hall effect may be substantial. As was shown by Berger [58], the Kohler rule does not hold in side jump scattering: indeed, the curves $E_X/\rho = f(H/\rho)$ for various temperatures are not identical (see Fig. 9 in [3]) in the alloy with $c_{Fe} = 32.2$ at%.

The "non-classical" contribution to the extraordinary Hall effect should be proportional to the square of the electrical resistivity. The function $R_s/\rho = f(\rho)$ was measured over a wide range of temperatures, both in fields greater than H_{cr} and less than H_{cr}; the electrical resistivity value at zero magnetic field was used for $H < H_{cr}$, while the electrical resistivity at $H > H_{cr}$ was used for $H > H_{cr}$ (see Fig. 8 in [3]). However, the term quadratic in electrical resistivity is also accompanied by a linear term, which indicates that classical asymmetrical scattering is also present.

Theoretical papers on the temperature dependence of the extraordinary Hall coefficient in ferromagnetic alloys [59, 60] for the case where the current carriers are scattered by phonons and the vibrational motion of the impurity atoms is taken into account indicate that the extraordinary Hall coefficient is a linear/quadratic function of temperature:

$$R_s^F = R_s^0 + AT + BT^2 \qquad (17)$$

where A and B are constants and R_s^0 is the "residual Hall coefficient" ($T \to 0$). This function also holds for $T > T_C$ (i. e., in the paramagnetic state).

In the case where the current carriers are scattered by magnons, the temperature dependence of the extraordinary Hall coefficient takes the following form [61]:

$$R_s^m = R_s^0 + B'T^2 + C'T^{3/2}, \qquad (18)$$

where B' and C' are constants, and R_s is independent of temperature for $T > T_C$. Neither of these functions is satisfied for $H > H_{cr}$ (see Fig. 10 in [3]).

For $T < T_1$ and $H < H_{cr}$ and $T > T_1$, the function $R_s(T)$ is better-described by the linear/quadratic function in (17), but with a negative coefficient in front of the linear term.

The fact that the coefficient R_p is constant for weak magnetic fields indicates that the contribution from scattering on magnetic inhomogeneities is important for $T > T_C$. However, the absence of a well-defined anomaly in the vicinity of T_C indicates that the non-magnetic contributions to the extraordinary Hall effect predominate.

Finally, since R_s and R_0 in Pt–Fe alloys with $c_{Fe} \geq 30.5$ at% turned out to be of opposite sign in the ferromagnetic region $(T_2 > T > T_3)$, it may be inferred that the main contribution to R_s is associated with the internal spin–orbit coupling constant for the interaction between "magnetic" d electrons [62]. This case corresponds to the case of a narrow d band, where the transport processes are primarily regulated by the s electrons.

Thus, the temperature dependence of the extraordinary Hall effect is evidently determined by the d-electron scattering on phonons and the s-electron scattering on magnetic inhomogeneities; asymmetric side-jump-like scattering also makes a substantial contribution to the extraordinary Hall effect. Relatively strong magnetic fields will change the ratios of the contributions from the various scattering mechanisms.

3.3. THE HALL EFFECT IN ALLOYS WITH $c_{Fe} < 30.5$ at%

Figures 13a and 13b show the functions $E_X(H)$ for the alloys with 27 and 28 at% Fe, respectively. The $E_X(H)$ curves are characteristically nonlinear, with the nonlinearity being most pronounced at $T = 4.2$ K. Only for the alloy with $c_{Fe} = 24.8$ at% and $T > T_2$ (i. e., for temperatures in the paramagnetic region) do the functions E_X become linear.

Because of the nonlinearity in the isotherms $E_X(H)$ and $E_X/H = f(J/H)$, we were only able to determine the Hall coefficients in the magnetic ordering region for the alloy with 27 at% Fe at a single temperature (95 K):

for fields $H < H_{cr}$ $R_0 = -(2 \pm 1) \cdot 10^{-12}$ Ohm $\cdot$ cm/G; $R_s = (10 \pm 2) \cdot 10^{-9}$ Ohm $\cdot$ cm/G;
for fields $H > H_{cr}$ $R_0 = -(0.8 \pm 0.4) \cdot 10^{-12}$ Ohm $\cdot$ cm/G; $R_s = (4.7 \pm 1) \cdot 10^{-9}$ Ohm $\cdot$ cm/G.

To determine the Hall coefficients in the paramagnetic temperature region, we constructed the functions $E_X = f(J/H)$ (see [3]) at fixed magnetic fields of $H = 20$, 70, and 120 kOe. In the alloy with $c_{Fe} = 27$ at%, the relationship was only observed to be linear at $H = 20$ kOe, but linearity is maintained even below the Néel temperature, and even all the way down to $T \approx T_1$. The magnetization measurements at 11 and 20 kOe also show no deviations in the function χ^{-1} over this temperature range. In the alloy with 24.8 at% Fe, the function $E_X/H =$

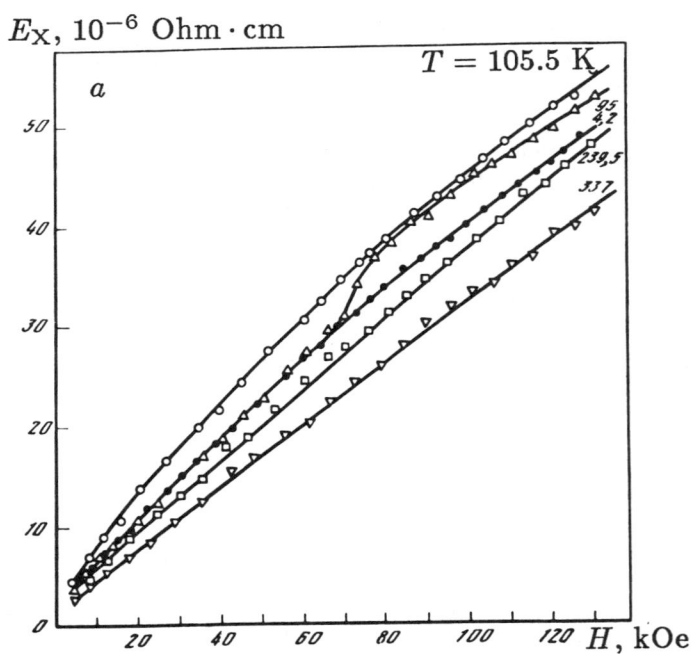

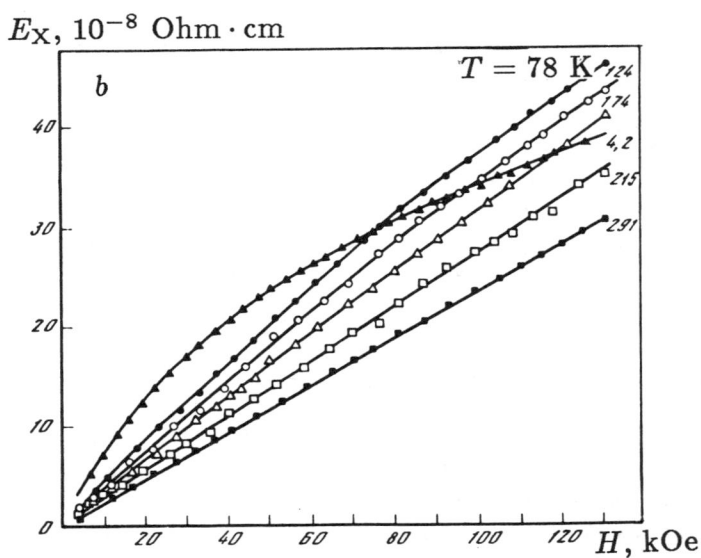

Figure 13   Hall EMF isotherms for iron–platinum alloys with 27 (a) and 24.8 at% Fe.

$f(J/H)$ is only linear above the Néel point. The linearity of the E_X and $J(H)$ curves for temperatures in the paramagnetic region means that the curves for $E_X/H = f(J/H)$ are independent of field strength.

The following table gives the values of the Hall coefficients in Pt–Fe alloys for temperatures in the paramagnetic region:

c_{Fe}, at%	24.8	27	32.2
R_0, 10^{-12} Ohm · cm/G	0.20 ± 0.05	2.4 ± 0.3	$-(6 \pm 1)$
R_p, 10^{-9} Ohm · cm/G	9 ± 1	2.4 ± 0.3	13.3 ± 0.5

In contrast to the case for alloys with $c_{Fe} > 30.5$ at%, R_0 and R_p are of identical sign and both positive in alloys with lower iron content. The sign of R_0 is a function of temperature in the alloy with $c_{Fe} = 27$ at%: negative for $T < T_1$ and positive at higher temperatures. The fact that the coefficient R_p is constant in the paramagnetic state indicates that the main contribution to the extraordinary Hall coefficient in the alloys with $c_{Fe} < 30.5$ at% comes from the spin scattering mechanism.

Quantitative estimates indicate that the contribution from the "side jump" mechanism is much smaller, especially at low temperatures. The Kohler rule is satisfied in the case of "classical" asymmetrical scattering, as we indeed observed for medium-strength fields in antiferromagnetic alloys of stoichiometric composition or with small deviations from stoichiometric composition.

3.4. HALL COEFFICIENTS AS A FUNCTION OF CONCENTRATION

The variation in R_p and R_0 in the paramagnetic region as a function of concentration is quite interesting (see above). The coefficient R_p is a non-monotonic function of composition, although it does remain positive: the minimum value of R_p occurs at $c_{Fe} = 27$ at%. The coefficient R_0 as a function of composition is also observed to have an extremum at this point, after which R_0 decreases and changes in sign from positive to negative at $c_{Fe} \approx 30$ at%.

Anisotropy in the scattering processes [63] and the electron distribution function [64] plays an important role in the behavior of the ordinary Hall coefficient in metals and alloys. These processes may lead to the appearance of maxima in the temperature dependence of R_0, but not to changes in sign as a function of temperature and composition.

The changes in the sign of R_0 may be regarded as being due to changes in the band structure which can lead to changes in the ratio of electron to hole mobility or the number of electrons and holes. This behavior of R_0 is apparently mainly due to changes in the alloy lattice parameters as iron is added to Pt_3Fe and lattice deformations resulting from changes in temperature and from magnetic ordering. The available experimental data can be taken as indicating that strong magnetic fields (or at least those resulting from magnetostrictive distortions) should also lead to changes in the band structure of the alloys.

We shall now briefly discuss the composition dependence of the extraordinary Hall coefficient. The theoretical work currently available consists of studies of the residual extraordinary Hall coefficient R_s^0 (rather than the coefficient in the paramagnetic temperature region) as a function of composition; as long as the coefficient is not a strong function of temperature (as is the case for Pt–Fe alloys with $c_{Fe} \geq 30.5$ at%), however, some of the qualitative characteristics in the behavior of $R_s^0(c)$ carry over to $R_p(c)$ as well.

Voloshinskii and Kovalenko [65] have shown that in completely disordered alloys with weak scattering, the sign of R_s^0 may depend on composition. Kondorskii, Vedyaev, and Granovskii [66, 67] (who used model approximations to take the band structure of Ni–Pd- and Fe–Pd-type alloys into account) showed that in the strong-scattering case, a "dip" should be expected in R_s^0 as a function of composition at some critical value $(y = y_{cr})$ for the concentration of component B in disordered alloys of the form $A_x B_y$ where components A and B are a strong ferromagnet and a paramagnetic substance, respectively $(Ni_x Pd_y)$.

A change in the sign of R_s^0 is to be expected in alloys of the form $Fe_x Pd_y$ (i. e., alloys between a weak ferromagnet A and a paramagnetic substance B) at $y = y_{cr,1}$, while a "dip" in R_s^0 as a function of composition without a change of sign is to be expected at $y = y_{cr,2}$.

The observed "dip" in R_p as a function of composition in Pt–Fe alloys can therefore be qualitatively explained within the framework of current theories of the extraordinary Hall effect which take the changes in alloy band structure as a function of composition into account.

Thus, the experimentally determined relationships for the Hall coefficients as a function of c_{Fe}, T, and H indicate that alloy band structure depends on composition and external thermodynamic parameters, with the band effects playing an increasingly important role at higher iron concentrations.

REFERENCES

1. Vinokurova, L. I., and V. Yu. Ivanov, "Electrical resistivity of ordered Pt–Fe alloys in the transition region from antiferromagnetism to ferromagnetism," Preprinti Fiz. Inst. Akad. Nauk (Lebedev Phys. Inst.), no. 21, 1982.

2. Vinokurova, L. I., and V. Yu. Ivanov, "An even galvanomagnetic effect (magnetoresistance) in ordered iron–platinum alloy monocrystals," Preprinti Fiz. Inst. Akad. Nauk (Lebedev Phys. Inst.), no. 22, 1982.

3. Vinokurova, L. I., and V. Yu. Ivanov, "The Hall effect in ordered monocrystalline Pt–Fe alloys," Preprinti Fiz. Inst. Akad. Nauk (Lebedev Phys. Inst.), no. 136, 1982.

4. Brandt, N. B., "Welding electrical conductors to samples," Pribory i Tekhnika Éksp., no. 2, pp. 138–140, 1956.

5. Kosichkin, Yu. V., "A study of the Fermi surface and magnetic breakdown in tellurium in extremely strong constant magnetic fields," Trudy Fiz. Inst. Akad. Nauk (Lebedev Phys. Inst.), vol. 67, pp. 8–49, 1973.

6. Tsiovkin, Yu. N., N. I. Kourov, and N. V. Volkenshtein, "Kinetic and magnetic properties of the alloy Pt_3Fe near the Néel point," Fiz. Tverd. Tela, vol. 20, pp. 940–942, 1978.

7. Cochrane, R. W., R. Harrig, J. O. Strom-Olsen, et al., "Structural manifestations in amorphous alloys: Resistance minima," Phys. Rev. Lett., vol. 35, pp. 676–679, 1975.

8. Haughton, R. V., M. R. Sarachik, and J. S. Kouvel, "Anomalous electrical resistivity and the existence of giant moments in NiCu alloys," Phys. Rev. Lett., vol. 25, pp. 238–239, 1970.

9. Al'tshuler, B. L., and A. G. Aronov, "On the theory of disordered metals and strongly alloyed semiconductors," Zh. Éksp. Teor. Fiz., vol. 77, pp. 2028–2044, 1979.

10. Loegel, B., "Magnetic transitions in the chromium–iron system," J. Phys. F: Metal Physics, vol. 5, pp. 497–505, 1975.

11. Ododo, J. C., and B. R. Coles, "The critical concentration for the onset of ferromagnetism in CuNi alloys," J. Phys. F: Metal Physics, vol. 7, pp. 2393–2400, 1977.

12. Berdyshev, A. A., and I. N. Vlasov, "On the electrical resistivity of an antiferromagnet," Fiz. Metallov i Metallovedenie, vol. 10, pp. 628–629, 1960.

13. Yamada, H, and S. Takada, "On the electrical resistivity of antiferromagnetic metals at low temperatures," Prog. Theor. Phys., vol. 52, pp. 1077–1093, 1974.

14. Kohgi, M., and Y. Ishikawa, "Magnetic excitation in a metallic antiferromagnet $FePt_3$," J. Phys. Soc. Japan, vol. 49, pp. 985–993, 1980.

15. Kourov, N. I., Yu. N. Tsiovkin, and N. V. Volkenshtein, "Low-temperature electrical resistivity of alloys with ferro- and antiferromagnetic interaction," Fiz. Nizkikh Temp., vol. 9, pp. 731–736, 1983.

16. Landau, L., and I. Pomeranchuk, "The properties of metals at very low temperatures," Zh. Éksp. Teor. Fiz., vol. 7, pp. 379–389, 1937.

17. Baber, W. G., "The contribution to the electrical resistance of metals from collisions between electrons," Proc. Roy. Soc. London A, vol. 158, pp. 383–396, 1937.

18. Kagan, Yu., and A. P. Zhernov, "The theory of electrical conductivity in metals with paramagnetic impurities," Zh. Éksp. Teor. Fiz., vol. 50, pp. 1107–1123, 1966.

18a. Kagan, Yu., and A. P. Zhernov, "The nonlinear composition dependence of the electrical resistivity in metals with impurities," Zh. Éksp. Teor. Fiz., vol. 60, pp. 1832–1844, 1971.

19. Masharov, S. I., "Electrical resistivity in disordered alloys," Fiz. Metallov i Metallovedenie, vol. 19, pp. 820–826, 1965.

19a. Masharov, S. I., "A multielectron theory for the electrical resistance of alloys," Fiz. Metallov i Metallovedenie, vol. 23, pp. 15–22, 1967.

20. Masharov, S. I, "Electrical resistivity in antiferromagnetic alloys," Phys. Status Solidi, vol. 21, pp. 747–753, 1967.

21. Ueda, K., "Electrical resistivity of antiferromagnetic metals," J. Phys. Soc. Japan, vol. 43, pp. 1497–1508, 1977.

22. Volkenshtein, N. V., V. P. Dyakina, and V. E. Startsev, "Scattering mechanisms of conduction electrons in transitions metals at low temperatures," Phys. Status Solidi B, vol. 57, pp. 9–42, 1973.

23. Murayama, S., and H. Nagasawa, "Magnetoresistance in antiferromagnetic α-Mn metal," J. Phys. Soc. Japan, vol. 43, pp. 1216–1223, 1977.

24. Kamada, M., N. Mori, and T. Mitsui, "Electrical resistivity near the critical boundary of antiferromagnet in $NiS_{2-x}Se_x$," J. Phys. C, Solid State Phys., vol. 10, pp. L643–L647, 1977.

25. Castaing, J., J. Danan, and M. Rieux, "Calorimetric and resistive investigation of the magnetic properties of CrB_2," Solid State Comm., vol. 10, pp. 563–565, 1972.

26. Ueda, K., and T. Moriya, "Contribution of s[pin fluctuation to the electrical and thermal resistivities of weakly and nearly ferromagnetic metals," J. Phys. Soc. Japan, vol. 39, pp. 605–615, 1975.

27. Ogawa, S., "Electrical resistivity of weak itinerant ferromagnet $ZrZn_2$," J. Phys. Soc. Japan, vol. 40, pp. 1007–1009, 1975.

28. Tarri, A., and B. R. Coles, "Electrical resistivity and the transition to ferromagnetism in the palladium–nickel alloys," J. Phys. F: Metal Physics, vol. 1, pp. L69–L71, 1971.

29. Simiyama, K., and G. M. Graham, "Magneto-volume effect in ordered Pt_3Fe alloy," Solid State Comm., vol. 19, pp. 241–243, 1976.

30. CRC Handbook of Chemistry and Physics, vol. 63, 1982/1983.

31. Crangle, J., and J. A. Show, "The range of stability of the superlattice Pt_3Fe," Phil. Mag., vol. 7, pp. 207–212, 1962.

32. Mott, N. F., "The resistance and thermoelectric properties of the transition metals," Proc. Roy. Soc. London A, vol. 156, pp. 368–382, 1936; "Electrons in transition metals," Adv. Phys., vol. 13, pp. 325–422, 1964.

33. Atoyan, A. M., A. F. Barabanov, and L. A. Maksimov, "Electrical resistivity of metals with unfilled f shells," Zh. Éksp. Teor. Fiz., vol. 74, pp. 2220–2233, 1978.

34. Allen, P. B., "Failure of the Boltzmann equation for nonlinear resistivity," Phys. Rev. Lett., vol. 37, pp. 1638–1641, 1976.

35. Mooij, J. H., "Electrical conduction in concentrated disordered transition metal alloys," Phys. Status Solidi A, vol. 17, pp. 521–530, 1973.

36. Kohler, M., "Zur magnetischen Widerstansänderung reiner Metalle," Ann. Phys., vol. 32, pp. 211–218, 1938.

37. Ziman, J. M., Electrons and Phonons, Oxford Univ. Press, Oxford, 1960 [Elektrony i Fonony, Izd. Inostr. Lit., Moscow, 1962].

38. Yamada, H., and S. Takada, "Magnetoresistance of antiferromagnetic metals due to s–d interaction," J. Phys. Soc. Japan, vol. 34, 51–57, 1973.

39. Usami, A. K., "Magnetoresistance in antiferromagnetic metals," J. Phys. Soc. Japan, vol. 45, pp. 466–475, 1978.

40. Novogrudnyi, V. M., and M. G. Fakidov, "Galvanomagnetic properties of $MnAu_3$," Fiz. Tverd. Tela, vol. 7, pp. 1095–1098, 1965.

41. Yosida, K., "Anomalous electrical resistivity and magnetoresistance due to an s–d interaction in Cu–Mn alloys," Phys. Rev., vol. 107, pp. 396–403, 1957.

42. Beal-Monod, M. T., and R. A. Wiener, "Negative magnetoresistivity in dilute alloys," Phys. Rev., vol. 170, pp. 552–559, 1968.

43. Zlatić, V., "Low temperature magnetoresistance of nearly magnetic transition metal-based alloys and actinides," J. Phys. F: Metal Phys., vol. 8, pp. 489–496, 1978.

44. Hamzic, A., and J. A. Campbell, "The magnetoresistance of PdFe alloys," J. Phys. F: Metal Phys., vol. 8, pp. 489–496, 1978.

45. Akhavan, M., H. A. Blackstead, and P. L. Donoho, "Magnetoresistance and field-induced phase transition in the helical antiferromagnetic state of dysprosium," Phys. Rev. B–Solid State, vol. 8, pp. 4258–4261, 1973.

46. Schinkel, C. J., R. Hartog, and F. H. A. M. Hochstenbach, "On the magnetic and electrical properties of nearly equiatomic ordered FeRh alloys," J. Phys. F: Metal Phys., vol. 4, pp. 1412–1422, 1974.

47. Adachi, K., M. Matsui, and Y. Omata, "Hall effect and magnetoresistance of $Co(S_xSe_{1-x})_2$, $0 \leq x \leq 1$," J. Phys. Soc. Japan, vol. 50, pp. 83–89, 1981.

48. Moriya, T., and K. Usami, "Coexistence of ferro- and antiferromagnetism and phase transitions in itinerant electron systems," Solid State Comm., vol. 23, pp. 935–938, 1977.

49. Kourov, N. I., Yu. N. Tsiovkin, and N. V. Volkenshtein, "Transverse magnetoresistance in atomically ordered $(Pd_xPt_{1-x})_3Fe$," Fiz. Tverd. Tela, vol. 23, pp. 1059–1064, 1981.

50. Belov, K. P., M. A. Belyanchikova, R. Z. Levitin, and S. A. Nikitin, Redkozemel'nye Ferromagnetiki i Antiferromagnetiki [Rare-Earth Ferromagnets and Antiferromagnets], Nauka, Moscow, 1965.

51. Döring, W., "Die Abhägigkeit des Widerstandes von Nickelkristallen von der Richtung der spontanen Magnetisierung," Ann. Phys., vol. 32, pp. 259–276, 1938.

52. Abdurakhmanov, A. A., Kineticheskie Effekty v Ferromagnitnykh Metallakh [Kinetic Effects in Ferromagnetic Metals], Izd. Rostovsk. Univ., Rostov-on-the-Don, 1973.

53. Pugh, E. M., "Hall effect and the magnetic properties of some ferromagnetic materials," Phys. Rev., vol. 36, pp. 1503–1511, 1930.

53a. Pugh, E. M., and T. W. Lippert, "Hall EMF and intensity of magnetization," Phys. Rev., vol. 42, pp. 709–713, 1932.

54. Kikoin, I. K., "Hall Effekt in Ni bein überschreiten des Curie Punktes," Phys. Zeitschr. UdSSR, vol. 9, pp. 1–12, 1936.

55. Kikoin, I. K., "The Hall effect in paramagnetic metals," Zh. Éksp. Teor. Fiz., vol. 10, pp. 1242–1247, 1940.

56. Vonsovskii, S. V., Magnetizm [Magnetism], Nauka, Moscow, 1971.

57. Berger, I., "Side-jump mechanism for the Hall effect of ferromagnets," Phys. Rev. B–Solid State, vol. 2, pp. 4559–4566, 1970.

58. Berger, I., "Hall effect of a compensated magnetic metal proportional to MB^2 in the high-field limit," Phys. Rev., vol. 177, pp. 790–792, 1969.

59. Voloshinskii, A. N., and N. V. Ryzhanova, "The extraordinary Hall effect in ferromagnetic alloys. I. Scattering on phonons," Fiz. Metallov i Metallovedenie, vol. 34, pp. 21–29, 1972.

60. Granovskii, A. B., "The extraordinary Hall effect in disordered alloys at high temperatures," Vestn. MGU. Fiz., Astron., no. 6, pp. 711–720, 1975.

61. Ryzhanova, N. V., and A. N. Voloshinskii, "The extraordinary Hall effect in ferromagnetic alloys. I. Scattering on magnons," Fiz. Metallov i Metallovedenie, vol. 35, pp. 269–276, 1973.

62. Irkhin, Yu. P., and Sh. Sh. Abel'skii, "Characteristics of kinetic phenomena and electron structure in transition metals," In: Trudy Mezhdunarodnogo Simpoziuma "Elektronnaya Struktura Perekhodnykh Metallov, Ikh Splavov i Soedinenii" [Proceedings of the International Symposium on the Electronic Structure of Transition Metals and Their Alloys and Compounds], Naukova Dumka, Kiev, pp. 94–99, 1974.

63. Kimura, H., and M. Shimuzu, "The effect of anisotropy in electron scattering on the hall coefficients for Pd metal and its alloys," J. Phys. Soc. Japan, vol. 20, pp. 770–778, 1965.

64. Kagan, Yu., and V. K. Flerov, "The theory of resistivity and magnetoresistance in metals at low temperatures," Zh. Éksp. Teor. Fiz., vol. 66, pp. 1374–1386, 1974.

65. Voloshinskii, A. N., and A. D. Kovalenko, "The theory of galvanomagnetic and magneto-optic phenomena in alloys," Fiz. Metallov i Metallovedenie, vol. 31, pp. 13–22, 1971.

66. Kondorskii, E. I., A. V. Vedyaev, and A. B. Granovskii, "The theory of the residual extraordinary Hall effect in disordered alloys. I; II," Fiz. Metallov i Metallovedenie, vol. 40, pp. 455–464; pp. 903–909, 1975.

67. Vedyaev, A. V., A. B. Granovskii, and E. I. Kondorskii, "Theory of the residual extraordinary Hall effect in disordered alloys. Weak scattering. II," Fiz. Metallov i Metallovedenie, vol. 40, pp. 688–694, 1975.

MAGNETIC PROPERTIES OF THE ALLOYS $Pt_3Mn_xFe_{1-x}$ AND $(Pd_xPt_{1-x})_3Fe$ IN STRONG MAGNETIC FIELDS

V. Yu. Ivanov, Yu. N. Tsiovkin, N. I. Kourov, and N. V. Volkenshtein

Abstract The magnetization of $Pt_3Mn_xFe_{1-x}$ and $(Pd_xPt_{1-x})_3Fe$ alloys was studied as a function of temperature and magnetic field strength (for fields of up to 150 kOe in the neighborhood of the critical concentration x_{cr} where most of the concentration-induced phase transition from the ferro- to the antiferromagnetic state occurs. We have also shown that the behavior of the magnetization at strong fields enables one to find a clear method of treating the magnetic state of the alloys in the neighborhood of x_{cr}. The effect of the magnetic and exchange anisotropy on the magnetic properties of the alloys was also discussed.

INTRODUCTION

Composition-induced phase transitions are observed in quasi-binary solid solutions of Pt_3Mn–Pt_3Fe and Pd_3Fe–Pt_3Fe in the ordered state (where the Pt and Pd atoms are located at the centers of the faces of the fcc lattice and the Fe and Mn atoms lie at the vertices of the cube) under changes in the concentrations of the components: in the cases of Pt_3Mn $(T_C = 400\,K)$ and Pd_3Fe $(T_C = 540\,K)$, the transition is into a ferromagnetic state, while in the case of Pt_3Fe $(T_N = 170\,K)$, the transition is into an antiferromagnetic (AFM) state.

An X-type magnetic state phase diagram with a quadruple critical point $T_C(x_{cr}) = T_N(x_{cr})$ (where x_{cr} is the transition concentration) was suggested in [1, 2] on the basis of neutron diffraction data. The X-type phase diagram means that the composition-induced AF–AFM phase transition in these systems occurs via an intermediate region with noncollinear magnetic structures.

Neither the quadruple critical point nor the second-order phase transition lines bounding the region with the noncollinear magnetic structures have been observed in studies of the electrical, thermoelectrical, galvanomagnetic, and thermal properties of these alloys [3–9]. Moreover, it turned out that $T_C(x_{cr}) > T_N(x_{cr})$. This led to the conclusion [3–9] that the composition-induced AFM–FM transition in

125

these alloys is a first-order phase transition. The lack of a common point with respect to the FM and AFM order parameters between the second-order phase transition lines led us to classify the magnetic state phase diagram as an F-type phase diagram [10].

Alloys with compositions in the transition region have several magnetic properties typical both of noncollinear structures and of interacting FM and AFM subsystems [11–13]. The most important of these is the fact that their magnetization σ peaks at a temperature corresponding to the appearance of AFM reflections in the neutron diffraction patterns. Two maxima and minima are observed in the temperature dependence of the differential magnetic susceptibility measured at $H \sim 15\,\mathrm{kOe}$. The initial magnetic susceptibility measured in variable magnetic fields has a characteristic maximum reminiscent of that in the spin glass state. However, the presence of coherent FM and AFM reflections in the neutron diffraction patterns of these alloys automatically rules out treating their magnetic states as spin-glass states.

Thermomagnetic processing procedures which include cooling the samples in a magnetic field from temperatures above T_N have a significant effect on the magnetization in alloys with $x \sim x_{\mathrm{cr}}$. In the alloy $(\mathrm{Pd_{0.53}Pt_{0.47}})_3$, for example, this processing leads to a shift in the hysteresis loop relative to the origin of the coordinate axis, while subsequent remagnetization has practically no effect on the form of the hysteresis loop. In the alloy $\mathrm{Pt_3Mn_{0.4}Fe_{0.6}}$, cooling in a field $H = 18\,\mathrm{kOe}$ leads to nearly a factor of two increase in σ over its value in the absence of thermomagnetic processing for $T < T_N$. This increase is easily undone by remagnetization, i. e., unusual thermomagnetic hysteresis effects are observed.

One characteristic trait of the alloys in the vicinity of x_{cr} is their high susceptibility at relatively high fields ($H \sim 15\,\mathrm{kOe}$), which may indicate that these alloys are far from magnetic saturation. The sharp increase in the anisotropy constants, specific electrical resistance, and the extraordinary Hall effect constants for the alloys with $x \sim x_{\mathrm{cr}}$ [3, 6, 11] should also be noted.

All of these features in the physical properties of the alloys under discussion for $H < 20\,\mathrm{kOe}$ indicate that the magnetization processes in them do not involve just the ordinary magnetic anisotropy. The exchange anisotropy due to the interaction between the FM and AFM subsystems evidently begins to play a role in the alloys with compositions in the transition region. Clarifying the role of both the magnetic and the exchange anisotropy will obviously require studies of the magnetic properties at stronger fields, as was in fact the aim of the present work.

Alloys in the immediate vicinity of the concentration x_{cr} are of natural interest in this respect. Two neutron-diffraction-verified samples with $x < x_{\mathrm{cr}}$ and $x < x_{\mathrm{cr}}$ each from the systems $\mathrm{Pt_3Mn_xFe_{1-x}}$ ($x = 0.3$ and 0.4) and $(\mathrm{Pd_xPt_{1-x}Fe}$ ($x = 0.33$ and 0.6) were used in the study. The σ measurements (at fields of up to $150\,\mathrm{kOe}$) were carried out at the "Solenoid" facility (Institute for General Physics, USSR Academy of Sciences) using a vibrating magnetometer.

EXPERIMENTAL RESULTS

The behavior of the magnetization is shown as a function of the external magnetic field and temperature in Figs. 1 and 2, respectively. As Fig. 1 indicates, the magnetization curves for the alloy $Pt_3Mn_{0.3}Fe_{0.7}$ change significantly with increasing temperature from 4.2 to 182 K. They become similar at higher temperatures, where they approach the Langevin forms for the function $\sigma(H)$. In the alloy $(Pd_{0.33}Pt_{0.67})_3Fe$, the function $\sigma(H)$ does not change significantly with increasing temperature; at fields $H > 40\,kOe$, the functions $\sigma(H)$ become practically identical and close to the Langevin functions.

As may be seen from Fig. 2, these alloys show a maximum in the function $\sigma(T)$ at strong and weak fields. This maximum is fairly sharp in the alloy $Pt_3Mn_{0.3}Fe_{0.7}$, and T_m, the temperature of the maximum, decreases rapidly with increasing temperature, with the shift in T_m being proportional to the field intensity (see inset in Fig. 2a). If we assume that it is the Néel point of the AFM matrix (which is where the maximum in $\sigma(T)$ is usually observed at small fields) that is changing, current ideas [14, pp. 60–80] indicate that the shift in T_m should be proportional to the square of the field, and should therefore not be so large. For $H > 60\,kOe$, the highly nonlinear function $\sigma(T)$ in this alloy above and below T_m becomes strikingly linear for $T > T_m$ over a broad temperature range.

At both weak and strong magnetic fields, the maxima in the functions $\sigma(T)$ are much less well-defined in the alloy $(Pd_{0.33}Pt_{0.67})_3Fe$ alloy than in the alloy $Pt_3Mn_{0.3}Fe_{0.7}$, while $\sigma(T)$ is a practically linear function of temperature in strong magnetic fields for $T < T_m$ and $T > T_m$.

Figure 3 shows the temperature dependence of the magnetic susceptibility χ measured in the alloys with $x < x_{cr}$ at strong and weak magnetic fields. One can see that the function $\chi(T)$ shows quite different behavior at strong fields. The low-temperature maximum in $\chi(T)$ becomes noticeably broader and is shifted to lower temperatures in this alloy. By $H \sim 120\,kOe$, the high-temperature maximum in $\chi(T)$ is almost entirely suppressed. Neither of the maxima in $\chi(T)$ is present in the alloy $(Pd_{0.33}Pt_{0.67})_3Fe$ for strong magnetic fields, while the susceptibility is a weak function of temperature on the interval 4.2–300 K; however, the value of χ is still quite large.

Figs. 4 and 5 show the behavior of the magnetic properties in the alloys with $x > x_{cr}$. Fig. 4 shows the magnetization curves $\sigma(H)$ obtained at various temperatures for the alloy $Pt_3Mn_{0.4}Fe_{0.6}$. The $\sigma(H)$ functions at $T < 100\,K$ for this alloy show a fairly broad induced-magnetization region. However, they saturate at fields $H > 90\,kOe$ for all of the temperatures studied.

Since thermomagnetic hysteresis is observed in this alloy upon cooling in magnetic fields less than the saturation field, the hysteresis loop for this alloy (a fragment of which is shown in the inset to Fig. 4) will be of interest. Fig. 4 shows that the magnetization hysteresis loop has a characteristic "constriction"; it is

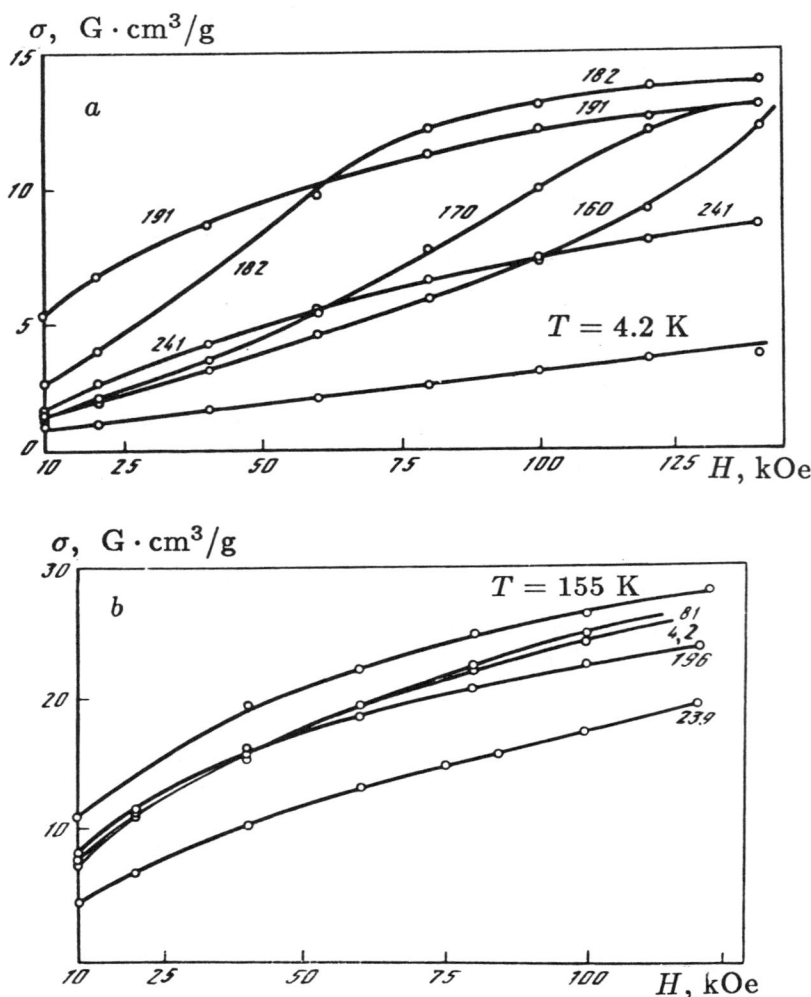

Figure 1 Magnetization as a function of external magnetic field at various temperatures. (a) $Pt_3Mn_{0.3}Fe_{0.7}$; (b) $(Pd_{0.33}Pt_{0.67})_3Fe$.

currently believed that this constriction indicates the state is unstable. A partial magnetization hysteresis loop is observed for $H < 50\,kOe$. However, magnetization values corresponding to the limiting hysteresis loop were observed when the sample was cooled in a magnetic field $H \sim 50\,kOe$ from temperatures greater than

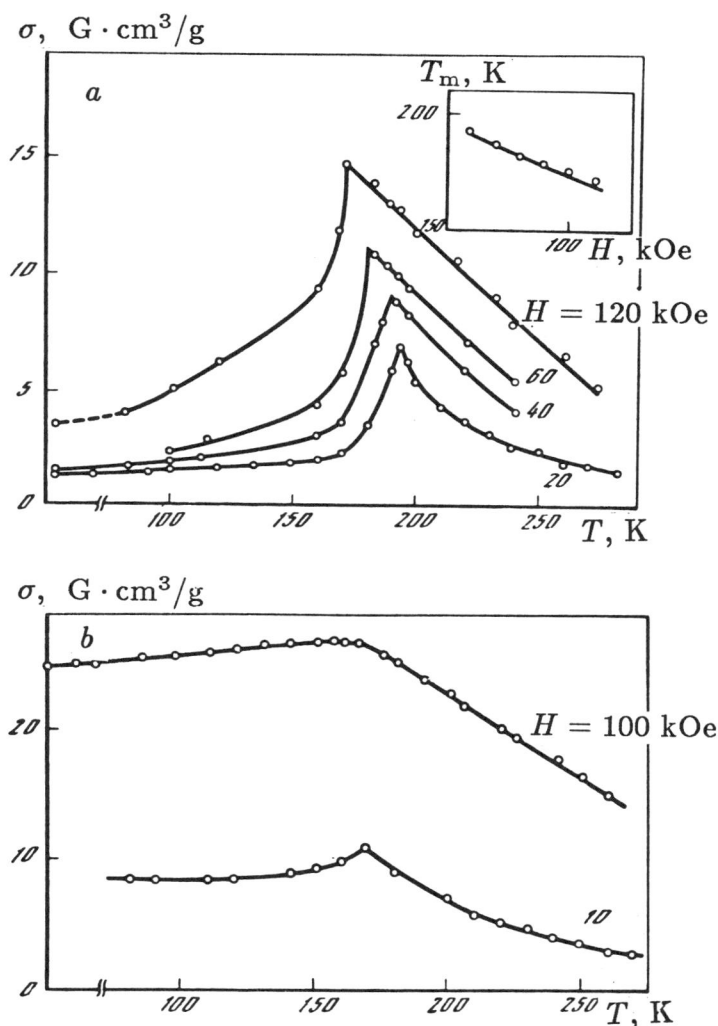

Figure 2 Magnetization as a function of temperature for various magnetic fields. (a) $Pt_3Mn_{0.3}Fe_{0.7}$; (b) $(Pd_{0.33}Pt_{0.67})_3Fe$. Inset: Temperature of the maximum in $\sigma(T)$ as a function of the external magnetic field.

T_N.

The magnetization measured for this alloy as a function of temperature in a weak magnetic field has a maximum at $T \sim 120\,K$ (see Fig. 5), which is approximately equal to T_N for the AFM subsystem. In fields $H \sim 90 \, \mathrm{kOe}$, this maximum is completely suppressed, and the function $\sigma(T)$ approaches the ferromagnetic form for $\sigma(T)$.

The magnetization functions for the alloy $(\mathrm{Pd}_{0.6}\mathrm{Pt}_{0.4})_3\mathrm{Fe}$ as a function of magnetic field strength have the characteristic form for ordinary ferromagnets, and they saturate at low magnetic fields ($\sim 20 \, \mathrm{kOe}$)—an indication that the degree of anisotropy in the magnetic field is small. The magnetization functions $\sigma(T)$ for this alloy as a function of temperature at $H \sim 100 \, \mathrm{kOe}$ are similar to those at

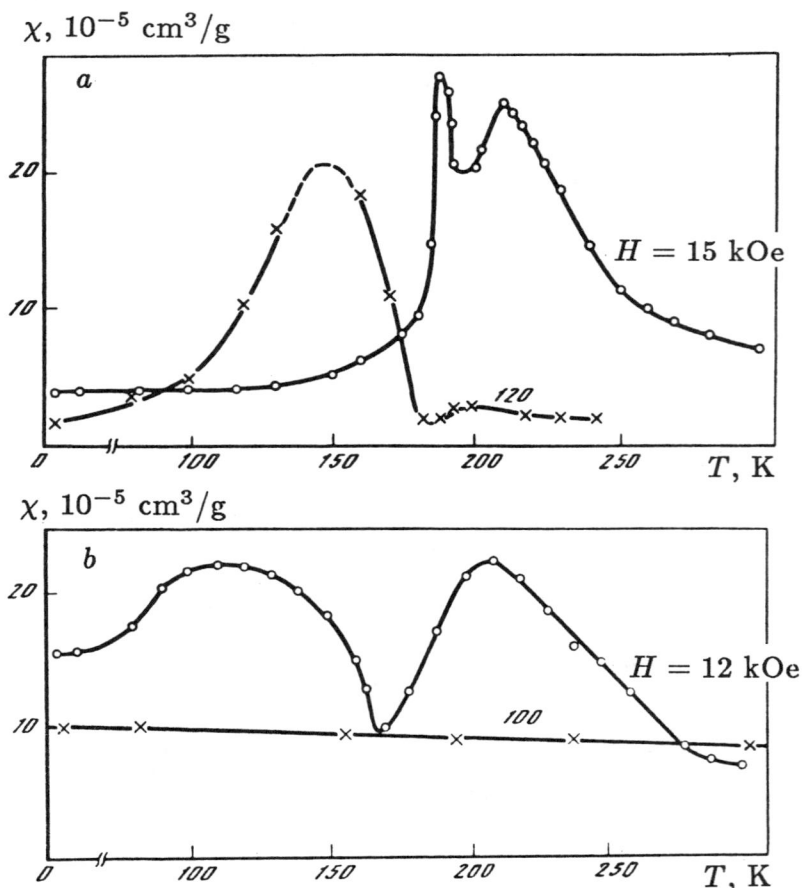

Figure 3   Magnetic susceptibility measured as a function of temperature for strong and weak fields. (a) $\mathrm{Pt}_3\mathrm{Mn}_{0.3}\mathrm{Fe}_{0.7}$; (b) $(\mathrm{Pd}_{0.33}\mathrm{Pt}_{0.67})_3\mathrm{Fe}$.

$H \sim 15\,\mathrm{kOe}$ (see inset in Fig. 5 and [11–13]). They are characterized by a maximum in $\sigma(T)$ at $T \sim 40\,\mathrm{K}$. However, the maximum in $\sigma(T)$ for this alloy cannot be explained in terms of a sharp increase in the field anisotropy for $T < T_\mathrm{m}$, since the magnetization curves are already saturated by $H \sim 20\,\mathrm{kOe}$.

DISCUSSION OF THE EXPERIMENTAL RESULTS

As is well known [1–10], the two alloy systems with competing exchange interaction studied here have identical magnetic phase diagrams. Analysis of the measurements presented above should, on the one hand, confirm the similarity between the magnetic states in these systems, and should reveal the difference between them due to the size and nature of the anisotropy forces on the other hand. A detailed study of the thermodynamics of this interacting mixture of FM and AFM subsystems was carried out in [15] without specifying a concrete microscopic model.

We shall discuss the magnetization in these alloys under the assumption that they have a nearly F-type magnetic phase diagram [10], with an AFM matrix containing FM clusters for $x > x_\mathrm{cr}$ and an FM matrix containing AFM clusters

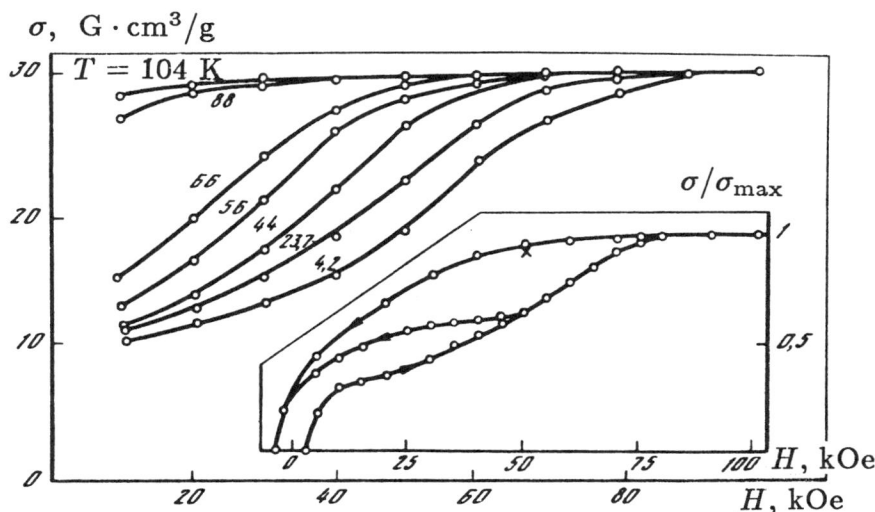

Figure 4   Magnetization as a function of external magnetic field at various temperatures for the alloy $\mathrm{Pt_3Mn_{0.4}Fe_{0.6}}$. Inset: Magnetization hysteresis loop for this alloy at $T = 4.2\,\mathrm{K}$; the cross indicates the $\sigma/\sigma_\mathrm{max}$ value obtained when the sample is cooled from $T = 300\,\mathrm{K}$ in a field $H = 50\,\mathrm{kOe}$.

for $x < x_{cr}$. The exchange interaction between the FM and AFM alloy subsystems (which leads to mutual ordering of their magnetic moments) is effective for $T \leq T_N < T_C$.

It is possible to show that in real alloys with $x < x_{cr}$, the FM–AFM exchange interaction can be described by an equation of the following form [16] when there is no phase boundary between the FM and AFM subsystems and finite FM clusters are formed by inversion of a moment in any one of the AF-matrix sublattices:

$$E_{int} \sim \mathbf{ML}, \tag{1}$$

where $\mathbf{L}$ is the magnetization of one of the sublattices in the AFM subsystem, $\mathbf{M} = \sum_i \mathbf{M}_i$, and $\mathbf{M}_i$ is the moment of the ith FM cluster. This is how the FM clusters in $Pt_3Mn_xFe_{1-x}$ alloys are formed.

The interaction energy between the finite AFM clusters and the FM matrix in the alloys under discussion is apparently given by a different expression for $x > x_{cr}$. The situation here will be analogous to that which obtains for antiferromagnets in an external magnetic field. If the field due to the FM matrix is parallel to the AFM cluster moments, the interaction energy will be equal to the sum of the interaction energies for each of the sublattices in the cluster:

$$E_{int} \sim \mathbf{M} \sum_i (\mathbf{L}_{1i} + \mathbf{L}_{2i}),$$

which is equal to zero for $|\mathbf{L}_{1i}| = |\mathbf{L}_{2i}|$.

If the molecular field due to the FM matrix is perpendicular to the AFM cluster moments, an induced moment will appear on the AFM clusters because the latter are not oriented parallel to the FM matrix field. The interaction energy between the FM and AFM subsystems is therefore of the form

$$E_{int} \sim \mathbf{M} \sum_i \mathbf{J}_i, \tag{2}$$

where $\mathbf{M}$ is the FM matrix moment, and $\mathbf{J}_i$ is the induced moment on the ith cluster in the molecular field of the FM matrix.

It is also of interest to consider the process by which finite FM clusters form within the AFM matrix in the $(Pd_xPt_{1-x})_3Fe$ alloys. As was shown in [13], the FM clusters in these alloys are due to the magnetic moments on the palladium atoms, i. e., the moments at the lattice points which are not part of the AFM cells. The latter is tantamount to planting a moment between the ferromagnetic sublattices. In this case, we once again have the analog of an antiferromagnet in an external field created by impurity moments, i. e., the FM–AFM interaction in the $(Pd_xPt_{1-x})_3Fe$ alloys can once again be described by equation (2).

We shall now analyze the behavior of the magnetization in the alloy systems under study, taking the concrete form obtained above for the interaction between the FM and AFM subsystems into account. First of all, we shall discuss the possible sources of the maximum in the functions $\sigma(T)$ for the $Pt_3Mn_xFe_{1-x}$ alloys.

According to (1), the exchange interaction between the FM and AF subsystems comes into play for $x > x_{cr}$ and $T \leq T_N$. If the interaction energy is greater than the energy of magnetic anisotropy in the two systems, each FM cluster moment will align itself along the antiferromagnetic axis of the matrix, with the result that the FM clusters will have a total mean moment of zero.

Two processes will be set into motion once an external magnetic field is applied: the matrix–cluster system as a whole will become magnetized, and the FM cluster moments will become detached from the AF subsystem sublattices, i. e., the FM–AFM exchange interaction will be overcome. The first of these processes obviously involves overcoming the matrix anisotropy field, which is $\sim 10\,kOe$ in these alloys (see [13]). The exchange anisotropy field is quite large here, and an external field $H \sim 150\,kOe$ is only sufficient to reorient the FM cluster moments in the vicinity of T_N. Thus, the maximum in the function $\sigma(T)$ at $T < T_N$ is due to the fact that the FM cluster moments are tied to the moments of the sublattices in the AFM matrix. The shift in T_m in the external magnetic field is clearly a result of overcoming the exchange anisotropy field.

The above discussion indicates that equation (2) should be used to describe

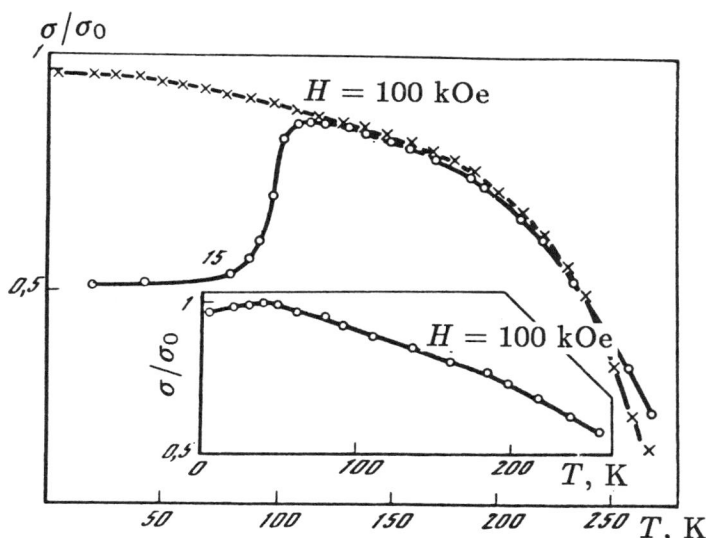

Figure 5   Magnetization of the alloy $Pt_3Mn_{0.4}Fe_{0.6}$ as a function of temperature at various magnetic fields. Inset: $(Pd_{0.6}Pt_{0.4})_3Fe$.

the interaction energy for $x > x_{cr}$. The energy in the exchange interaction can be assumed to be greater than the energy of magnetic anisotropy in each of the subsystems. We obviously have $\mathbf{J}_i \| \mathbf{M}$, i. e., $\mathbf{J}_i \perp \mathbf{L}$, in the ground state. At this point, the magnetization process will now mainly be determined by the magnetic anisotropy of interacting matrix–cluster subsystems.

This therefore implies that the maximum in the function $\sigma(T)$ at fields $H < 20$ kOe (less than the saturation field) is due to magnetic anisotropy forces. The lack of a singularity in the $\sigma(T)$ curve at $H > 90$ kOe indicates that the AFM clusters have low magnetization and do not make a significant contribution to the total magnetization of the alloy.

Thus, the behavior of the magnetization as a function of temperature and magnetic field in the $Pt_3Mn_xFe_{1-x}$ alloys enables us to sharply distinguish between the regions with $x < x_{cr}$ and $x > x_{cr}$, and is a direct consequence of the presence of an interacting mixture of FM and AFM subsystems for concentrations in the transition region.

It was shown above for the $(Pd_xPt_{1-x})_3Fe$ alloys that the magnetization is determined solely by the magnetic anisotropy energy. Since the latter is small in these alloys, the curves $\sigma(H)$ for the samples with $x < x_{cr}$ and $x > x_{cr}$ are already saturated at fields $H \sim 30$ kOe. However, this does not make it clear why a maximum is observed for fields $H \sim 100$ kOe (i. e., much greater than the magnetic anisotropy field). We believe that the answer to this question also depends on the concrete form of the FM–AFM interaction. If we assume that the ferromagnetic polarization of the AFM subsystem is sufficiently large, we then have a significant moment $\mathbf{J}_i$. In this case, the saturation magnetization is the sum of the magnetization due to the FM and AFM subsystems. The first of these is a monotonic function of temperature, while the second has a temperature dependence similar to that of a noncollinear antiferromagnet. The observed maximum in the resulting function $\sigma(T)$ for these alloys can thus be attributed to the noncollinear antiferromagnet part of the magnetization as a function of temperature [17]. The large value of the total magnetization in a $(Pd_{0.33}Pt_{0.67})_3Fe$ sample whose palladium concentration differs significantly from x_{cr} relative to that in an $Pt_3Mn_{0.3}Fe_{0.7}$ alloy with nearly-critical composition [12, 13] is experimental proof of the presence of a significant moment $\mathbf{J}_i$ in these alloys.

The behavior of the susceptibility in these alloys also confirms that the interaction between the FM and AFM systems is as suggested above. The FM cluster moments in the alloys with $x < x_{cr}$ become mutually ordered at some $T_C' > T_N$. This temperature corresponds to the high-temperature maximum in the function $\chi(T)$ at weak fields. The FM and AFM subsystems begin to interact at $T \leq T_N$, and leads to reorientation of the magnetic moments in the FM clusters and to their becoming correlated with the antiferromagnetic axis of the matrix. This orientation phase transition is accompanied by the appearance of a low-temperature maximum in the $\chi(T)$ curves for $T \lesssim T_N$ and magnetic fields less than the anisotropy field. At higher fields (i. e., greater than the characteristic anisotropy fields), the susceptibility should be a weak function of temperature. This is confirmed by the fact that at strong fields, χ is a weak function of temperature in the alloy $(Pd_{0.33}Pt_{0.67})_3Fe$.

The experimental results obtained here lead to the unambiguous conclusion that the exchange anisotropy due to the interaction between the FM and AFM subsystems plays a substantial role (along with the magnetic anisotropy) in the structure of the magnetic state, and thus also plays a role in the behavior of the physical properties in alloys with competing exchange interaction along with the magnetic anisotropy.

REFERENCES

1. Dubinin, S. F., A. P. Vokhmyanin, V. V. Kelarev, et al., "The magnetic structure of disordered solid solutions of d transition metals with exchange interactions of various sign based on the example of $Mn_{1-c}Fe_cPt_3$," Fiz. Metallov i Metallovedenie, vol. 48, pp. 764–773, 1979.

2. Kelarev, V. V., A. P. Kozlov, A. P. Vokhmyanin, et al., "Magnetic transformations in ordered $Fe(Pt_\alpha Pd_{1-\alpha})$ alloys as a function of temperature," Fiz. Metallov i Metallovedenie, vol. 34, pp. 977–981, 1972.

3. Kourov, N. I., Yu. N. Tsiovkin, and N. V. Volkenshtein, "Electrical resistance of the $Pt_3Mn_xFe_{1-x}$ alloys with interacting magnetic order parameters," Fiz. Metallov i Metallovedenie, vol. 55, pp. 955–959, 1983.

4. Kourov, N. I., Yu. N. Tsiovkin, and N. V. Volkenshtein, "Thermoelectromotive force in atom-ordered $(Pd_xPt_{1-x})Fe$ alloys," Fiz. Tverd. Tela, vol. 21, pp. 1511–1514.

5. Podgornykh, S. M., N. I. Kourov, Yu. N. Tsiovkin, et al., "Galvanomagnetic properties of the $Pt_3Mn_xFe_{1-x}$ alloys with mixed exchange interaction," Fiz. Metallov i Metallovedenie, vol. 58, pp. 265–270, 1984.

6. Kourov, N. I., Yu. N. Tsiovkin, and N. V. Volkenshtein, "The Hall effect in $(Pd_xPt_{1-x})_3Fe$ alloys with mixed exchange interaction," Fiz. Metallov i Metallovedenie, vol. 54, pp. 678–684, 1982.

7. Kourov, N. I., Yu. N. Tsiovkin, and N. V. Volkenshtein, "Transverse magnetoresistance in atom-ordered $(Pd_xPt_{1-x})_3Fe$ alloys," Fiz. Tverd. Tela, vol. 23, pp. 1059–1064, 1981.

8. Kourov, N. I., S. M. Podgornykh, Yu. N. Tsiovkin, et al., "Heat capacity of the $Pt_3Mn_xFe_{1-x}$ alloys," Zh. Éksp. Teor. Fiz., vol. 79, pp. 1921–1926, 1980.

9. Kourov, N. I., Tsiovkin, Yu. N., S. M. Podgornykh, et al., "Heat capacity of alloys with interacting magnetic order parameters," Zh. Éksp. Teor. Fiz., vol. 83, pp. 662–667, 1982.

10. Podgornykh, S. M., N. I. Kourov, Yu. N. Tsiovkin, et al., "The magnetic phase diagram of the Pt_3Mn-Pt_3Fe and Pd_3Fe-Pt_3Fe alloys," In: Vsesoyuznaya konferentsiya po fizike magnitnykh yavlenii (9 sent. 1983): Tezisy dokladov (All-Union Conference on the Physics of Magnetic Phenomena: Summaries of Papers), Tula, pp. 274–275, 1983.

11. Kadomatsu, H., "The magnetic properties of the ordered alloys Fe (Pd_xPt_{1-x})," J. Sci. Hiroshima Univ. A, vol. 37, pp. 141–145, 1973.

12. Tsiovkin, Yu. N., N. I. Kourov, and N. V. Volkenshtein, "Magnetic properties of ordered solid Pt_3Mn-Pt_3Fe solutions," Fiz. Metallov i Metallovedenie, vol. 58, pp. 1137–1143, 1984.

13. Tsiovkin, Yu. N., N. I. Kourov, and N. V. Volkenshtein, "Magnetic state of the $(Pd_xPt_{1-x})_3Fe$ alloys with mixed exchange interaction," Fiz. Tverd. Tela, vol. 23, pp. 2614–2620, 1981.

14. Borovik–Romanov, A. S., Antiferromagnetizm i ferrity (Antiferromagnetism and Ferrites), Izd. Akad. Nauk SSSR, Moscow, 1962.

15. Vlasov, K. B., and A. I. Mitsek, "The thermodynamic theory of materials of materials which can be either ferromagnetic or antiferromagnetic. I. Magnetization processes," Fiz. Metallov i Metallovedenie, vol. 14, pp. 487–497, "II. Temperature dependence of the parameters which determine the magnetic state and the shape of the magnetization curve," Fiz. Metallov i Metallovedenie, vol. 14, pp. 498–502, 1962.

16. Meiklejohn, W. H., and C. P. Bean, "New magnetic anisotropy," Phys. Rev., vol. 102, pp. 1413–1414, 1956; vol. 105, pp. 904–913, 1957.

17. Nagaev, É. L., "Anomalous magnetic structures and phase transitions in non-Heisenberg magnetic substances," Usp. Fiz. Nauk, vol. 136, pp. 61–103, 1982.

FERMI SURFACES OF MAGNETIC
TRANSITION METALS UNDER PRESSURE

A. G. Gapotchenko, E. S. Itskevich, and É. T. Kulatov

Abstract A method was developed for measuring the de Haas–van Alphen effect at pressures of up to 30 kbar. The results of a study of the effect of pressure on the Fermi surfaces of ferromagnetic iron, cobalt, and nickel and antiferromagnetic chromium were also presented. A model was proposed to explain the experimentally observed signs and values of the relative changes in the extremal cross sections of various pieces of the Fermi surfaces in these magnetically ordered metals under pressure.

INTRODUCTION

One of the most interesting properties of the transition metals is the presence of a magnetically-ordered state. The practical need for magnetic materials, the problem of obtaining new materials which have atom-by-atom magnetic ordering under certain external conditions, and the rapid development of the theory of condensed matter in recent years have necessitated deeper theoretical and experimental study of the electronic structure of magnetically-ordered transition metals, since the electron structure determines almost all of the properties of the metals. These studies are characterized by the use of powerful computers on the one hand, and the use of experimental techniques such as the measurement of quantum oscillation effects, neutron diffraction measurements, and experiments in ferromagnetic and antiferromagnetic optics and magneto-optics, on the other.

The main parameters used to describe the magnetically-ordered state in metals include the following: the magnetic moment per atom, the temperature of the transition into the magnetic state (T_C or T_N), the exchange splitting between spin subbands (which is generally a function of the wave vector **k**), the density of states in the spin subbands, and the Fermi surface.

Verifying the adequacy of theoretical models for the magnetic ordering phenomenon requires that the properties of the metals be measured as a function of interatomic separation, i. e., that they be measured at various pressures. Unfortunately, most experimental methods are extremely difficult or even impossible to use under conditions of high pressure. In this respect, the de Haas–van Alphen (DHVA) effect provides a unique opportunity for direct observation of the electron structure near the Fermi surface (FS) and its variation with pressure even to the point where the topology of the Fermi surface is substantially altered. The DHVA effect enables one to identify and very accurately measure signals from different pieces of the Fermi surface, and is quite sensitive to extrinsic thermodynamic properties such as pressure and temperature. The relative change in DHVA oscillation frequency can be measured to an accuracy of order 10^{-4}, and this means that the shift in the energy levels relative to the Fermi level can be measured to comparable accuracy.

1. MEASUREMENT METHOD

1.1. THE DE HAAS–VAN ALPHEN EFFECT

Experimental methods in solid state physics based on quantum oscillatory effects provide direct information on the Fermi surface and the band structure of metals near the Fermi surface. Many properties of metals such as the magnetic moment, magnetoresistance, speed of ultrasound, absorption of ultrasound, and thermoelectromotive force are oscillatory and depend on the applied field, but these quantum oscillatory effects only occur under special conditions, i. e., at low temperatures, in fairly strong magnetic fields, or in very pure monocrystals of the material under study.

Oscillations in the magnetic susceptibility are known as the DHVA effect. This effect is actively used to study the electronic properties of metals, ordered intermetallic compounds, and dilute substitution alloys.

A great deal of basic theoretical research has been carried out on the DHVA effect, and this problem can now be considered completely solved. The most detailed theory of the DHVA effect is that constructed by Lifshits and Kosevich [1].

Lifshits and Kosevich obtained the following expression (in the spherical coordinates $\hat{\mathbf{B}}$, θ, ϕ) for the oscillatory part of the magnetic moment, $\mathbf{M}_{\text{osc}}$:

$$\mathbf{M}_{\text{osc}} = \sum_j \left\{ -F_j, \frac{\partial F_j}{\partial \theta}, \frac{1}{\sin \theta} \frac{\partial F_j}{\partial \phi} \right\} \sum_{r=1}^{\infty} Z_{j,r}^{\text{osc}} \sin \left(r \frac{2\pi F_j}{B} + \gamma \right) \qquad (1)$$

where γ is the phase factor (equal to $1/2$ for a free electron gas); r is the number of the harmonic; $\mathbf{B}$ is the magnetic induction; F_j, G, is the DHVA frequency, which is related to the area of the jth extremal cross section of the Fermi surface A_j, a. u.$^{-2}$, via the equation $F_j = \hbar c A_j / 2\pi e$, or, in practical units,

$$F_j = 3.741 \cdot 10^8 A_j. \tag{2}$$

Note that Kittel [2] has provided a basis for generalizing this method to a system of interacting magnetically-ordered electrons.

The amplitude of the rth harmonic for the jth extremal cross section of the Fermi surface at temperature T takes the following form:

$$Z_{j,r}^{\text{osc}} = \sqrt{\frac{2}{\pi}} \left(\frac{e}{\hbar}\right)^{3/2} k_B \left|\frac{\partial^2 A_j}{\partial k_z^2}\right|^{-1/2} \frac{T}{r^{1/2}} \frac{\exp\{\alpha r m_j^* T_{D,j}/B\}}{\sinh\{\alpha r m_j^* T/B\}} \cos\left(\frac{r g_j m_j^b \pi}{2}\right), \tag{3}$$

where m_j^* is the effective electron mass, m_j^b is the band electron mass, g_j is the orbital g-factor, $T_{D,j}$ is the Dingle temperature for the jth cross section,

$$\alpha = \frac{2\pi k_B m_e}{e\hbar},$$

m_e is the free electron mass, k_z is the component of the electron momentum parallel to the field, and k_B is the Boltzmann constant.

We shall now present a brief discussion of the individual terms in the expression for M_{osc} and determine what information they contain on the band structure and the Fermi surface.

1. Most of the variation in M_{osc} is due to the $\sin(2\pi F_j/B)$ term. The quantity $2\pi F_j/B$ determines the phase of the oscillation, while the ratio F_j/B is (to the nearest integer) the quantum number n of the oscillation. The DHVA frequency in metals usually lies between 10^6 and 10^9 Hz, so that the corresponding phases range from 200π to $200\,000\pi$ for fields of order $10\,\text{kG}$. Thus, one can record a large number of complete oscillation cycles (100–300), determine the DVHA frequency F_j to high accuracy, and then calculate (using (2)) the area of the extremal cross section A_j of the Fermi surface by varying the magnetic field over a sufficiently wide range.

2. The term $\partial^2 A_j / \partial k_z^2$ is related to the local curvature of the Fermi surface in the vicinity of the jth cross section; the term $\cos(r g_j m_j^b \pi/2)$ is related to the Zeeman splitting of the Landau levels. Both contributions can be measured experimentally from the absolute values of the amplitudes at the DHVA frequency and the amplitude ratios of the various harmonics, respectively.

3. The term

$$\frac{T}{\sinh(r\alpha m_j^* T/B)} \approx 2T \exp\left(-\frac{r\alpha m_j^* T}{B}\right) \tag{4}$$

describes the temperature dependence of $Z_{j,r}^{\text{osc}}$. The effective electron mass m_j^* can be determined by measuring the amplitude of the DHVA frequency as a function of temperature in a constant magnetic field.

The lifetimes of the electronic states near the Fermi level are shorter in a real metal due to the presence of lattice defects, inhomogeneities, impurities, etc. This leads to broadening of the Landau levels, and the term

$$\exp(-r\alpha m_j^* T_{D,j}/B) \tag{5}$$

describes the impact of this on the DHVA effect. The Dingle temperature [3] $T_{D,j}$ introduced here is related to the electron relaxation time τ (which is independent of the temperature T) via the equation

$$T_{D,j} = \frac{\hbar^2}{2\pi} \frac{1}{k_B \langle \tau \rangle_j}, \tag{6}$$

where $\langle \tau \rangle$ is the relaxation time averaged over the jth orbit on the Fermi surface. If the effective mass m_j^* is known, the Dingle temperature can be determined from the amplitude $Z_{j,r}^{\text{osc}}$ as a function of magnetic field at constant temperature.

The second summation in equation (1) is carried out over all harmonics of the fundamental DHVA frequency; however, the experimental oscillation curves are very nearly sinusoidal, which means that the amplitudes of the higher-order harmonics in expansion (1) decrease rapidly with increasing temperature and decreasing relaxation time τ, and $\mathbf{M}_{\text{osc}}$ as a function of $\mathbf{B}$ can be described to sufficient accuracy using the first harmonic alone:

$$\mathbf{M}_{\text{osc}} = \sum_j z_j^{\text{osc}} \sin\left(\frac{2\pi}{B} F_j + \gamma\right). \tag{7}$$

1.2. APPARATUS USED TO MEASURE THE DHVA EFFECT BY THE MODULATION METHOD

The modulation method suggested by Shoenberg [4] is currently the most widely used for measuring the DHVA effect. The theory and apparatus for the modulation method have been described in detail by Stark and Windmiller [5]. We shall now briefly discuss the idea behind the method. We shall restrict ourselves to a discussion of a nonferromagnetic sample in an external magnetic field $\mathbf{H}_0$ located inside a system consisting of a detector and modulation coil. Superposition of a modulated field $\mathbf{h}_m(t) = \mathbf{h}_0 \cos \omega t$ on the slowly varying external magnetic field H_0 leads to a substantial increase in the induced voltage on the receiving coil. The modulation field should satisfy two requirements: first, the amplitude of modulation $|\mathbf{h}_0|$ should be small compared with $|\mathbf{H}_0|$, and secondly, the frequency of modulation should be low enough that the thickness of the skin layer in the sample will be much larger than the size of the sample. Equation (7) can then be written in the following form (for simplicity, we shall consider only the jth component of the DVHA effect):

$$M_j^{\text{osc}}(t) = Z_j^{\text{osc}}(\mathbf{H}_0, T, T_D) \sin\left[\frac{2\pi F_j}{|\mathbf{H}_0 + \mathbf{h}_m(t)|} + \gamma\right]. \tag{8}$$

Using the condition $|\mathbf{h}_0|/|\mathbf{H}_0| \ll 1$, we find, to first order in this small parameter,

$$M_j^{\text{osc}}(t) = Z_j^{\text{osc}}(\mathbf{H}_0, T, T_D) \sin\left[\left(\frac{2\pi F_j}{|\mathbf{H}_0|}\right) - \Lambda_j \sin \omega t\right], \tag{9}$$

where

$$\Lambda_j = 2\pi F_j |\mathbf{h}_0|/|\mathbf{H}_0|^2.$$

Standard Fourier analysis of equation (9) yields:

$$M_j^{\text{osc}}(t) = 2 \sum_{n=1}^{\infty} n\omega J_n(\Lambda_j) Z_j^{\text{osc}}(\mathbf{H}_0, T, T_D) \sin\left(\frac{2\pi F_j}{|\mathbf{H}_0|} - \frac{\pi n}{2} + \gamma_j\right) \sin(n\omega t)$$

$$+ J_0(\Lambda_j) Z_j^{\text{osc}}(\mathbf{H}_0, T, T_D) \sin\left(\frac{2\pi F_j}{|\mathbf{H}_0|} + \gamma_j\right). \tag{10}$$

The induced voltage in the coil is given by dM_j^{osc}/dt, i. e.,

$$V_f(t) = \frac{dM_j^{\text{osc}}}{dt} = N_L \Omega Z_j^{\text{osc}}(\mathbf{H}_0, T, T_D)$$

$$\times \sum_{n=1}^{\infty} n\omega \sin\left(\frac{2\pi F_j}{|\mathbf{H}_0|} - \frac{\pi n}{2} + \gamma_j\right) J_n(\Lambda_j) \sin(n\omega t), \tag{11}$$

and contains all harmonics of the modulation frequency. In (11), N_L is the number of turns per unit length in the detector coil, and Ω is the sample volume. Also note that we have been discussing the case of longitudinal modulation, i. e., $\mathbf{h}_m(t)$ parallel to $\mathbf{H}_0$.

If we observe only the nth harmonic, the voltage at the output of the phase detector will be given by

$$V^{n\omega} = \sum_j n\omega N_L Z_j^{\text{osc}}(\mathbf{H}_0, T, T_D) J_n(\Lambda_j) \sin\left(\frac{2\pi F_j}{|\mathbf{H}_0|} - \frac{\pi n}{2} + \gamma_j\right). \tag{12}$$

Thus, M^{osc} as a function of $\mathbf{H}_0$ appears to be modulated in amplitude at the secondary frequency $n\omega$ characteristic of the DHVA terms, whose amplitudes are modified by the nth order Bessel function. The presence of the $J_n(\Lambda_j)$ in (12) leads to a substantial improvement in the instrumental selectivity: by selecting $\mathbf{h}_0$ and $\mathbf{H}_0$ appropriately, we can obtain an argument $\Lambda_j = 2\pi F_j h_0/H_0^2 = 3.05$, i. e., close to the first maximum in the Bessel function $J_2(3.05) = 0.5$ (in our experiments, we usually detected the second harmonic of the modulation frequency $n = 2$), while $J_2(\Lambda_{j'})$ will be much smaller for the other frequencies. This allowed us to select out the DHVA frequencies alone and suppress the others.

Frequency discrimination is effective for detecting useable signals at the higher harmonics of the modulation frequency, where $J_n(\Lambda_j)$ is a stronger function of h_0. This the case of overmodulation, where $\Lambda_j \geq 2\pi$, i. e., $h_0 \geq H_0^2/F_j = \Delta H$, and the modulation amplitude $(|\mathbf{h}_0|)$ is greater than or equal to the oscillation period of the field (ΔH) itself [6].

A superconducting solenoid 100 mm in length, with an inside diameter of 20 mm and an outside diameter of 80 mm was used to generate the magnetic field. The solenoid was calibrated using the nuclear magnetic resonance technique developed by Teplinskii [7]. This same method was also used to measure the homogeneity of the field at the center of the solenoid $(1 \cdot 10^{-4}\,\mathrm{mm}^{-1})$ and the reproducibility of the field, i. e., the degree to which $\mathbf{H}$ remained constant at a given current from experiment to experiment $(2 \cdot 10^{-4})$.

The maximum field that could be produced by the superconducting solenoid was 80 kOe at $T = 1.5$ K, while the coupling factor between the field and solenoid current $k = 2.5815\,\mathrm{kOe/A}$. The DVHA effect measurements were carried out at liquid-helium temperatures ranging from 4.2 to 1.5 K.

Figure 1 shows the general view of the system of measuring coils located in the high-pressure chamber. The internal channel of the Derlin housing attached to the outside of obturator (6) is 2.0 mm in diameter; the outside diameter is 4.3 mm, and the housing walls have a thickness of 0.15 mm. The sensor coil (1), compensation coil (2), and modulation coils (3) each consist of 600 turns of water-resistant enamel wire 0.03 mm in diameter. The sensor and compensation coils are connected in opposition, so that the external induction signal is reduced by a factor of approximately 100. The amplitude of the modulation field generated by the modulation coil can be varied from 0 to 60 kOe. Placing the measuring coils inside the high-pressure region enabled us to obtain a high coupling coefficient

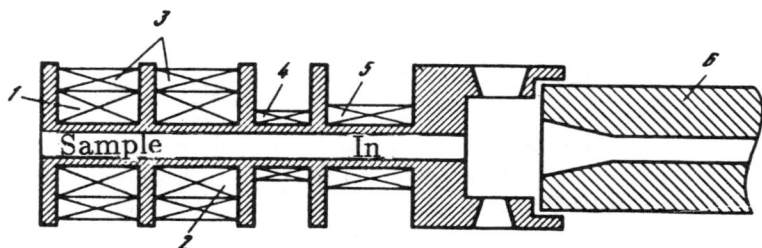

Figure 1   System of measuring coils and sensors for measuring the DHVA effect and the pressure in the fixed-pressure chamber.

between the sample and the coils and to detect changes in the magnetic moment in small crystals of not very high quality.

A block diagram of the instrument is shown in Figure 2. The sample was located inside receiving coil (2). The EMF due to the modulation field from coil (3) at frequency ω was mutually absorbed in the detector and compensation coils. The EMF induced in the sensor coil by the magnetic moment of the sample was then fed into the measuring system. Since the detector coil has an output impedance of approximately 5 Ohm and while the U2-6 selective amplifier has an input impedance of 10 kOhm, the sensor coil had to be matched to the amplifier circuit. This was achieved by inserting a step-up transformer between the detector coil and the U2-6 amplifier.

To avoid overloading the U2-6 amplifier with the noncompensated part of the modulation voltage at frequency ω, the signal was passed through a double T filter (which cut off this frequency ω) after the matching transformer.

After amplification of the voltage by the U2-6 amplifier and synchronous detection of the frequency 2ω, the signal was then fed into the Y coordinate of a chart recorder. The X coordinate of the recorder was simultaneously fed a voltage

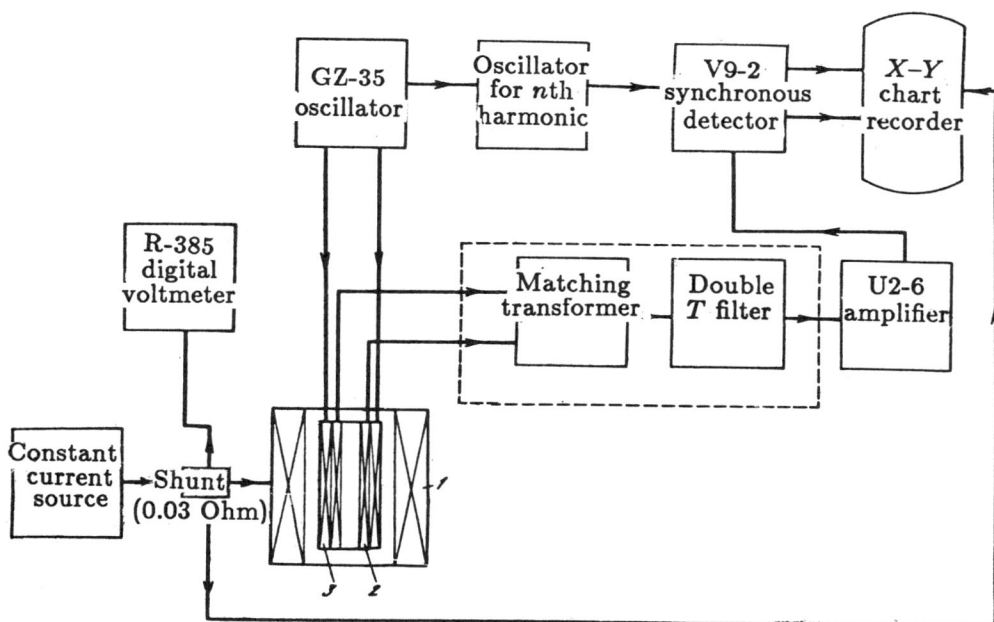

Figure 2   Block diagram of the apparatus for measuring the DHVA effect.

proportional to the current (field) in solenoid (1) from the shunt (0.03 Ohm). This same voltage was fed into an R-385 digital voltmeter whose readings were recorded on the chart recorder paper (in the form of edge marks). A G3-35 oscillator (9) fed a 185-Hz voltage to the modulation coils (3).

1.3. THE GENERATION AND MEASUREMENT OF PRESSURE FOR STUDYING THE DE HAAS–VAN ALPHEN EFFECT

In the present work, we are using the fixed-pressure chambers suggested by Itskevich [8]. These chambers have the advantage that it is relatively simple to create pressures of up to 15 kbar by compressing a liquid medium (transformer oil–kerosene) at room temperature. The self-regulating nature of the chamber makes it simple to carry out measurements at various pressures with the sample in a fixed orientation.

A general view of the fixed-pressure chamber used in our experiments is shown in Fig. 4. The body (1), with an outside diameter of 18 mm and an inside diameter of 4.5 mm, was made of heat-treated beryllium bronze with an HRC hardness of 38–40. The piston (2) and obturator (3) screws, the obturator (4), and the plunger (5) were made of the same bronze. The piston (6) and mushroom-shaped tip (7) were made of non-magnetic age-hardened 40KhNYu steel heat-processed to an HRC hardness of 58–60. The sample was located inside the sensor coils (8). This system enables one to obtain stable pressures of up to 12 kbar at helium temperatures without disruption or changes in the dimensions of the parts. The "plug-type" obturator (with a packing consisting of a set of copper and beryllium-bronze rings) works well up to pressures ~ 19 kbar, where the beryllium bronze rapidly loses strength. The piston packing consisted of a conical beryllium-bronze ring, two copper rings, and a polyurethane ring. The piston packing is replaced after each cycle of measurements (the polyurethane ring shows practically no wear, and is reused many times).

The process of obtaining pressure at room temperature was monitored using a sensor consisting of manganin wire wrapped around one section of the sensor coil (item (4) in Fig. 1). After setting the pressure and fixing the position of the piston using the screw, the chamber (attached to a measuring sonde) was placed in a cryostat.

Since a decrease in pressure Δp (Δp ranges between 2 and 3.5 kbar) takes place as the compression mixture hardens, the true pressure at helium temperatures was measured using the temperature of the transition to superconductivity in indium as a function of pressure [9, 10]. The error in the pressure amounted to approximately 0.2 kbar. The transition to superconductivity was recorded using the sensor coil (item (5) in Fig. 1).

In order to measure the impact of compression on the DHVA effect, we require that the deviations from hydrostatic equilibrium in the fixed-pressure chamber

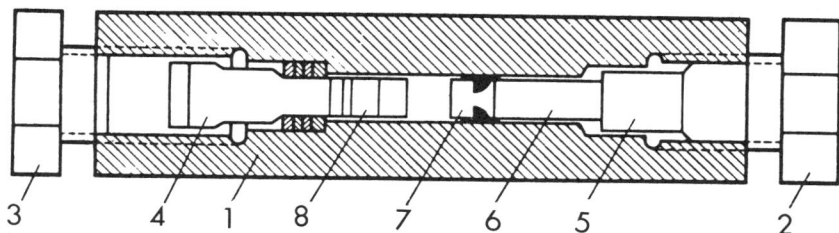

Figure 3   The fixed-pressure chamber.

be fairly small. As is well known, true hydrostatic equilibrium at liquid-helium temperatures can only be obtained using liquid helium, and then only on the narrow pressure range from 0 to 140 bar. When other methods (solid helium, fixed-pressure chambers) are used, the degree of hydrostatic equilibrium is usu-ally estimated from the extent to which pressure gradients distort the measure-ment results. The criterion used for this is to compare the results of measurements obtained using various methods, including those made under true hydrostatic con-ditions, which, as pointed out by Itskevich, et al. [11], should, to within the errors of measurement, be identical. From this point of view, fixed-pressure chambers are quite suitable for measuring the de Haas–van Alphen effect under pressure.

Note, however, that the oil–kerosene mixture used as a working medium de-velops pressure gradients and shear stresses (which may damage the samples and lead to spurious results) upon solidification at pressures greater than 15 kbar. The measurements were therefore carried out at pressures less than or equal to 12 kbar.

A further step has recently been taken in the development of low-temperature chambers for studying quantum oscillation effects at pressures of up to 30–35 kbar. The most promising and simplest to use turned out to be one having a beryllium-bronze cylindrical body with a press-fitted non-magnetic steel bushing and hardened-alloy compound piston [12]. In addition to the usual pressure-adjustment screw, this chamber is equipped with a lockscrew between the adjustment screw and bushing and an extruded ring between the piston itself and the mushroom packing, while the obturator is conical in form. The usable portion of the chamber is 6 mm in diameter. Oil–pentane mixtures are used as the pressure-transmitting medium within the chamber.

1.4. MEASUREMENT OF THE STRESS DERIVATIVES FOR VARIOUS CROSS SECTIONS OF THE FERMI SURFACE

Two methods were used to measure the stress derivatives of the Fermi surface. In the first method, the set of experimental values $F(p)$ was approximated using the method of least squares by a linear function $F = F_0 + ap$. The stress derivative in this case is given by the equation

$$d \ln F / dp = a / F_0. \tag{13}$$

In the second method, the stress derivative was determined via the following expression from the size of the phase shift in the oscillations with pressure:

$$\frac{d \ln F}{dp} = \frac{B}{F} \frac{\Delta \phi}{\Delta p} \frac{1}{2\pi}, \tag{14}$$

where $\Delta \phi / 2\pi$ is the relative shift in phase for a change in pressure of Δp. This shift in phase was recorded using the shift in the point marking the magnetic field relative to its initial position in the oscillations. The experimental function $\phi(p)$ was approximated by the straight line $\phi = \phi_0 + ap$, while the resulting shift was approximated by $\Delta \phi = a \Delta p - \phi_0$, where ϕ_0 is the phase of the oscillation at zero pressure.

The experimentally measured values of the stress derivative are usually small $(\simeq 10^{-3} \, \text{kbar}^{-1})$, so that we then have the problem of keeping the experiment as accurate as possible. All random and systematic errors must be kept to a minimum. Possible systematic errors include the errors in the calibration of the measuring instruments and in the calibration of the field values.

The random errors are primarily due to the lack of reproducibility of the field, and may be reduced by choosing a definite procedure for the measurements. Random errors may also result if the sample deviates from its initial orientation. This may lead to variations in F which are not a function of pressure.

The estimated relative measurement accuracy for the oscillation frequency in our experiments is $\sim 8 \cdot 10^{-4}$.

2. THE EFFECT OF PRESSURE ON THE FERMI SURFACE IN FERROMAGNETIC TRANSITION METALS AND IN RHENIUM

2.1. TOPOLOGY OF THE FERMI SURFACE AT ATMOSPHERIC PRESSURE

As is well known, the conditions for the appearance of conduction-electron ferromagnetism, i. e., band ferromagnetism, were first formulated by Stoner in 1931 [13]. The Stoner model suggests that the energy bands are each split into two subbands of opposite electron spin, with the size of this splitting being proportional to the magnetization $m_s = n_\uparrow - n_\downarrow$, where n_σ is the number of electrons with spin $\sigma = (\uparrow, \downarrow)$. In the ferromagnetic state, the number of electrons with spin oriented parallel to the direction of magnetization is greater than the number of electrons with spin antiparallel to the direction of magnetization. We shall hereinafter refer to these subbands as majority electron spin subbands and minority electron spin subbands, respectively.

The Fermi surface for iron at atmospheric pressure has been studied by the DHVA method in several papers. The most important and pioneering of these papers (in the respect that it was the first reported observation of the DHVA effect in a ferromagnetic metal) was [14]. Unusually complete measurements of Fermi surface cross sections for iron in spherical monocrystals were carried out in [15, 16]; even sections of order the size of the Brillouin zone were observed. The Fermi surface of iron is currently believed to consist of the following regions: in the majority electron spin subband, we have a large electron surface centered on the point Γ, two hole pockets centered on the point H, and a system of hole protuberances along the line $H-N$; in the minority electron spin subband, we have an electron "jack" centered on the point Γ bordering the hole octahedron at the point H along the line Δ, a hole ellipsoid at N, and electron balls along the line Δ. The Fermi surfaces for iron in the two spin subbands are shown in Fig. 4.

The Fermi surface for nickel at atmospheric pressure has also been studied fairly completely [17, 18]. In the majority electron spin subband, the Fermi surface for nickel consists of a single sp surface which is open in the $\Gamma-L$ direction because of the necks at point L. The Fermi surface in the majority electron spin subband very closely resembles the Fermi surface of copper; however, the neck is much smaller in diameter in copper than in nickel. In the minority electron spin subband, we have two large d-like surfaces centered on the point Γ and a hole ellipsoid at the point X. Note that theory predicts a second hole ellipsoid at the point X that has not been observed in any of the DHVA effect experiments. Figure 5 shows the Fermi surface of nickel in both spin subbands.

The least-studied of the three ferromagnetic $3d$ metals is cobalt. This is due to the low quality of cobalt monocrystals because of the fcc–hcp structural transition at $T = 390$ K, where internal stresses lead to partitioning of the monocrystals into blocks. Papers in which quantum oscillation effects are used to measure some of the extremal cross sections of the Fermi surface in cobalt have nevertheless recently begun to appear.

To within the experimental error (which is mainly associated with the value of the internal field), the experimental results in [19, 20] confirm the existence of five DHVA frequencies between 1 and 14 MGs. Moreover, Coleman, et al. [21] have observed de Haas–Shubnikov oscillations at frequencies of 1.05, 3.57, and 11.7 MGs in filamentary monocrystalline cobalt crystals.

Several theoretical papers [22–24] yield contradictory results on the topology of the Fermi surface of cobalt in the hcp phase. A new model has recently been

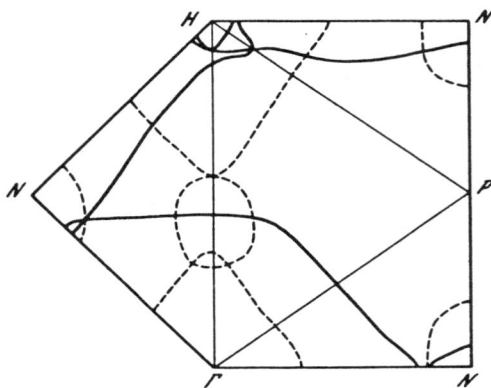

Figure 4   Fermi surface for ferromagnetic iron in the majority electron spin subband (solid lines) and the minority spin electron subband (dashed lines).

proposed for the Fermi surface in ferromagnetic hcp cobalt [25]. This model differs in several respects from models of the Fermi surface calculated in earlier papers. This model indicates that the majority spin subband Fermi surface consists of two large sheets: a multiply-connected electron surface with necks centered on the point Γ, a hole monster with a protuberance extending along the line $L-A$; deformed hole balls; and dumbbells and an ellipsoid at the point L (Fig. 6, solid lines). Moreover, the intersection between the ellipsoid and the protuberances of the hole monster leads to the formation of an α ball which was not present in any of the earlier calculations.

Research on the Fermi surface in rhenium at atmospheric pressure has now been under way for a number of years. Several DHVA frequencies (which have been ascribed to various extremal cross sections of the Fermi surface) agree in shape and size with the Fermi surface of hcp rhenium theoretically calculated by Matheiss [30]. Closed hole pockets (resembling an ellipsoid, dumbbell, and deformed ball, respectively) centered on the point L were observed in bands five, six, and seven. These surfaces are quite similar to the three Fermi surfaces recovered from measurements of the DHVA effect in [26]. Note that the Fermi surface for rhenium has the same topological features near the point L as the Fermi surface for cobalt in the minority electron spin subband at the point L. This fact will be used below when we discuss models for the additive contributions to the pressure derivative in a ferromagnet.

Also, the calculation in [30] indicates that the Fermi surface of rhenium consists of an open electron surface similar to a corrugated cylinder with an opening along

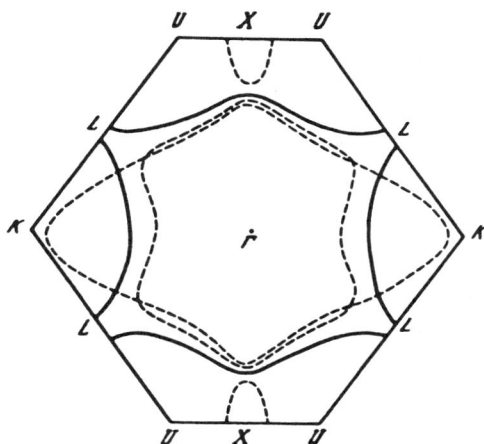

Figure 5   Fermi surface of ferromagnetic nickel. Solid lines: majority spin subband; dashed lines: minority spin subband.

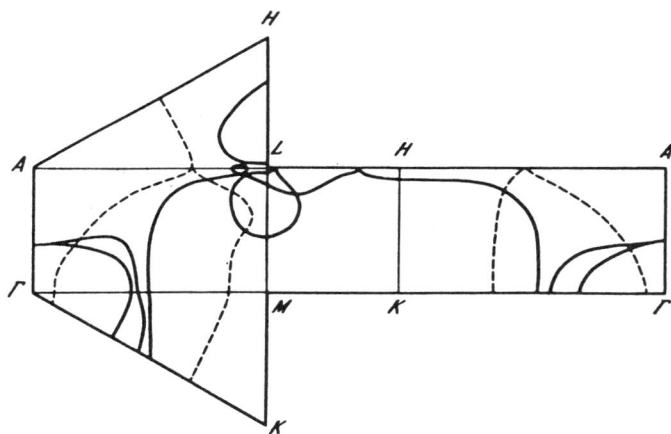

Figure 6   Fermi surface for ferromagnetic hexagonal cobalt.

the [0001] axis in band eight; a small, closed hole pocket at the point Γ; and, an electron surface with as yet undetermined topology in band nine. There are, however, indications that this surface has toroidal topology, i. e., a ring consisting of six spheres connected by necks.

2.2. THE SAMPLES

The nickel, iron, cobalt, and chromium samples were cut from monocrystals using an electric arc cutter. After cutting, the samples were etched in a hot saline acid solution to remove surface stresses. The crystallographic axes of the samples were determined using X-ray diffraction.

The chromium samples were cut in the shape of parallepipeds with dimensions of $1.3 \times 1.3 \times 3$ mm. Samples of two types were used in the DHVA effect measurements: in the former, the sample axis was identical to the [100] axis, while it was identical to the [110] axis in the latter.[1] The ratio of the electrical resistivity of the chromium samples at room temperature to that at liquid-helium temperatures was 130.

The cobalt samples were also cut in the shape of parallelepipeds, with dimensions of $1.0 \times 1.0 \times 4$ mm. The ratio $\rho_{300 \text{K}}/\rho_{4.2 \text{K}}$ was approximately 150. However, as the X-ray measurements indicated, the cobalt monocrystals were turned out to be divided into blocks.

The $1 \times 1 \times 3$ mm nickel samples were cut from monocrystalline bars having a ratio $\rho_{300 \text{K}}/\rho_{4.2 \text{K}} \approx 3000$. The measurements were carried out on type $\langle 111 \rangle$, $\langle 100 \rangle$, and $\langle 112 \rangle$ samples.

The nickel and cobalt crystals were obtained from the USSR Academy of Sciences Institute for Solid State Physics. The chromium monocrystals were kindly provided to us by T. I. Kostin (Moscow State University), and the rhenium samples were obtained from the Institute for Metal Physics at the USSR Academy of Sciences Ural Science Center.

The iron samples (in the form of two monocrystalline spheres 1.5 mm in diameter) were obtained from the Max Planck Institute in Stuttgart (FRG). The ratio $\rho_{300 \text{K}}/\rho_{4.2 \text{K}}$ was approximately 500 in these samples.

2.3. EFFECT OF PRESSURE ON THE FERMI SURFACES OF IRON, NICKEL, COBALT, AND RHENIUM

We do not know of any data on the Fermi surface of iron under pressure. All that can be mentioned in this connection is the paper by Angadi, Fawcett, and Rasolt [31], who studied the effect of stretching on the magnetoresistance of Fe crystal filaments (whiskers).

The effects of pressures of up to 11 kbar on the oscillation frequencies were measured with the samples in three orientations relative to the external magnetic

---

[1]For brevity, we shall hereinafter call the samples with a certain crystallographic axis such as the $\langle 100 \rangle$ axis, etc., aligned with the long side of the parallelepiped type-$\langle 100 \rangle$, etc., samples.

field: $\mathbf{H}\|[100]$, $[110]$, and $[111]$. The internal field $\mathbf{H}_{int}$ in the iron was equal to 14 660 G. The variations in the F_c-type DHVA frequencies, i. e., in the singlet field F_s from the small lens cross section (one of the smallest regions in the minority spin subband Fermi surface of iron) and the modulation frequency F_s of the doublet frequency F_d associated with the longitudinal section of the lens [32] were studied for all three crystal orientations mentioned above. At atmospheric pressure, these frequencies turned out to be as follows: $F_s = 3.8\,\text{MGs}$ and $F_d = 4.2\,\text{MGs}$, in agreement with the values obtained in [33]. A single frequency was recorded with $\mathbf{H}\|[111]$, since the extremal cross sections of all six lenses in the (111) plane are identical by symmetry. In the orientation $\mathbf{H}\|[110]$, a beat pattern between the frequencies F_s and F_d due to the longitudinal cross sections of the two lenses centered on the [001] axis and to the noncentral cross sections of the remaining four lenses, respectively, was observed.

Figure 7 shows the pattern of DHVA effect oscillations in iron for $\mathbf{H}\|[110]$. The figure indicates that pressure has a substantial effect on the positions of the beat frequencies. The large errors for $\mathbf{H}\|[100]$ are due to the oscillations from the "lens" being mixed with oscillations from the "jack" collar (one of the small pieces in the minority spin subband Fermi surface of iron).

The experimental values of the pressure derivatives (in $10^{-3}\,\text{kbar}^{-1}$) for the singlet and doublet lens cross sections S_s and S_d are given below:

Field orientation	[100]	[111]	[110]
PD_s	8.5 ± 2	8.5 ± 0.5	9.5 ± 1.0
PD_d	8.5 ± 2	8.0 ± 0.5	9.5 ± 1.0

The pressure derivatives turned out to be quite large—nearly a factor of 20 larger than the compressibility of iron, which is $-5.2 \cdot 10^{-4}\,\text{kbar}^{-1}$. Unlike the pressure derivative, the effective masses m^* and Dingle temperature T_D are not affected by pressure (to within the experimental error).

Oscillations with frequencies corresponding to the minimal neck cross section in the majority spin subband, i. e., the piece of the Fermi surface centered on point L of the Brillouin zone, are observed in type-$\langle 111 \rangle$ nickel samples for magnetic fields of 10–80 kOe at all pressures (up to 11 kbar). In fields of 60–80 kOe, these oscillations become superposed on the oscillations associated with the noncentral (111) cross section of the hole ellipsoid centered on the point X, but in the other spin subband, i. e., the minority spin subband.

Oscillations corresponding to the (100) cross section of the same ellipsoid were observed in the type-$\langle 100 \rangle$ samples for fields between 35 and 80 kOe.

When the field orientation $\mathbf{H}\|[112]$, the oscillations were observed to start at $H = 35$ kOe; these oscillations were associated with the skew cross section of the neck in the (112) plane. By $H \approx 45$ kOe, hole ellipsoid oscillations begin to appear superposed on the observed oscillation pattern.

The induction in the samples was determined by minimizing the root-mean-square deviation of the quantity $\sqrt{\sum (\Delta n)^2 / N}$ in the least-squares approximation of the experimental function $n = F\Delta H / (H_1 + H_{int})(H_2 + H_{int})$ by a straight line for various values of the internal field H_{int} as a parameter (where n is the number of the oscillation, F is the DHVA frequency, and H_1 and H_2 are the values of the solenoid field at the beginning and end of a train of oscillations). The results indicated that $H_{int} \approx 6400\,\text{G}$.

All of the oscillation frequencies recorded at zero pressure are in agreement with the results in [18] to within the accuracy with which the axes of the sample could be adjusted relative to the magnetic field.

All of the oscillation frequencies were found to be linear functions of pressure on the interval 0–11 kbar, and the pressure derivatives turned out to be fairly small (Table 1).

Our data imply that the hole ellipsoid becomes more anisotropic under pressure: the minor cross section remains nearly constant, and the major cross section and the neck cross section each increase in size upon compression.

The pressure derivative determined for the neck is in agreement with the results in [34], which were obtained using the solid-helium high-pressure technique. The pressure derivative for the ellipsoid is quite small, less than the limiting accuracy of our measurements. This is what apparently explains the difference between our results and those in [34], where the value of the pressure derivative was also small, but of the opposite sign: $(1 \pm 0.2) \cdot 10^{-4}\,\text{kbar}^{-1}$.

As far as we know, Anderson, et al. [35], who measured the variation in the extremal cross section β (attributed to the neck centered on the point Γ in the majority electron spin band for a field oriented along the [0001] direction, i. e., in the basis plane) as a function of pressure, are the only ones to have studied the

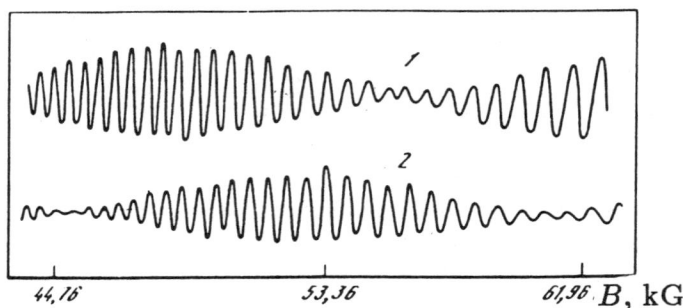

Figure 7 Magnetic susceptibility oscillations in iron for H‖[110] at two pressures: (1) 7 kbar; (2) 10 kbar.

effect of pressure on the Fermi surface of cobalt:

$$d \ln F_\beta / dp = (-2.2 \pm 0.3) \cdot 10^{-3} \, \text{kbar}^{-1}.$$

The pressure derivative obtained in [34] is unusually large in absolute value for a transition metal with relatively low compressibility ($k_T = 5.13 \cdot 10^{-4} \, \text{kbar}^{-1}$ at 4.2 K [36, pp. 279–313]). Moreover, the Fermi surface cross section usually increases with pressure in the transition metals; in this case, however, the pressure derivative is negative. The single frequency in zirconium and the extremely-low-frequency oscillations in gadolinium are two apparent exceptions to this rule. The sign of the pressure derivative cannot be explained by anisotropy of the lattice deformations, since the variation in c/a with pressure is quite small:

$$d \ln(c/a)/dp = 6.17 \cdot 10^{-6} \, \text{kbar}^{-1}.$$

In order to obtain new information on the Fermi surface in cobalt, we carried out several experiments with the purpose of studying the DHVA effect under pressure. Oscillations were recorded at fields higher than 30 kOe (the results in [37] indicate that the difference in direction between the external field and the induction in the sample is no greater than 1°).

Oscillations with frequency $F_\alpha = 1.06 \, \text{MGs}$ were recorded at atmospheric pressure with the magnetic field $\mathbf{H} \| [0001]$.

Because of the small number of oscillations recorded, the pressure derivative of the α cross section could be determined to the required accuracy only by measuring the phase shift under pressure. The period of oscillation for the α cross section in cobalt at $B \simeq 50 \, \text{kG}$ was 2 kG, and the phase shift turned out to be equal to π at $p = 10 \, \text{kbar}$, so that the error in determining the position of the oscillation due to nonreproducibility of the field was quite small. Equation (14) was used to calculate the phase shift.

Measurements were carried out on two samples of cobalt, and the value of the pressure derivative of the α cross section in these samples for $\mathbf{H} \| [0001]$ turned out

Table 1. Experimental Values of the Pressure Derivative for Various Pieces of the Fermi Surface in Nickel

Piece of Fermi Surface	Cross Sectional Plane	Pressure Derivative, $10^{-4} \, \text{kbar}$	Piece of Fermi Surface	Cross Sectional Plane	Pressure Derivative, $10^{-4} \, \text{kbar}$
Neck at L	(111) (112)	8.0 ± 2.0 6.6 ± 2.5	Ellipsoid at X	(111) (112) (100)	6.6 ± 2.5 1.5 ± 0.8 -0.8 ± 0.8

Table 2. Experimental Values of the
Pressure Derivative for the Ellipsoid,
Dumbbell, and Ball Pieces of the Fermi
Surface in Rhenium

Fermi Surface Cross Section	F, MGs	Pressure Derivative $10^{-4}\,\mathrm{kbar}^{-1}$
Ellipsoid, h_5	0.78	8.9 ± 3.0
Dumbbell, h_6	14.5	4.5 ± 2.0
Ball, h_7	64.3	2.5 ± 1.0

to be

$$d \ln S_\alpha / dp = (-1.4 \pm 0.3) \cdot 10^{-3}\,\mathrm{kbar}^{-1}.$$

The pressure derivative for the α cross section turned out to be large and negative, like that for the β cross section. This unusual behavior of the Fermi surface in cobalt will be discussed below in the light of calculations of the Fermi surface in cobalt both under normal conditions and under pressure.

The DHVA effect was measured in high-purity rhenium crystals ($\rho_{300\,\mathrm{K}}/\rho_{4.2\,\mathrm{K}} = 5500$) at pressures of up to 10 kbar. As was already mentioned above, the aim of the study was to obtain information on the pressure derivative of various pieces of the Fermi surface, since rhenium is an electronic analog of the minority spin subband in cobalt (rhenium has 7 valence electrons per atom, while the minority spin subband in cobalt has 7.44). Moreover, rhenium, with an hcp crystal lattice like cobalt, has a nearly ideal ratio of c/a (1.633), which remains practically constant in both metals for pressures of up to 10 kbar.

At atmospheric pressure, oscillations of frequency 0.78 MGs (according to the model of the Fermi surface of rhenium proposed by Matthiess, these are due to the minimal cross section of the band five hole ellipsoid (h_5) centered on the point L [30]) were observed to begin at magnetic fields of order 6 kG. By ~ 22 kG, oscillations of frequency 14.5 MGs due to the band six dumbbell-like cross sections of the Fermi surface (also centered on the point L (h_6)) were observed to be superposed on the oscillations of frequency $F(h_5)$. And, finally, a third frequency $F(h_7) = 64.3$ MGs (associated with the deformed hole ball in band seven which is also centered on the point L) was observed at fields greater than ~ 57 kG.

The pressure derivatives of these frequencies are given in Table 2. The quoted errors for the pressure derivatives are equal to the magnitude of the dispersion. As one can see from the table, the pressure derivatives are quite small for the cross sections being measured, as is typical of low-compressibility ($k_T = -2.69 \cdot 10^{-4}$ kbar^{-1} [38]) metals without magnetic order.

Measurements of the effect of pressure on the effective mass m^* associated with ellipsoidal cross section h_5 (which is equal to $0.1m_0$ at atmospheric pressure) indicated that m^* remains constant, i. e., $d\ln m^*/dp = 0$, on the pressure interval 0–10 kbar. We were not able to successfully measure the pressure derivatives of the dumbbell and ball effective masses, since the amplitudes of the oscillations from these pieces of the Fermi surface varied quite erratically with pressure, which evidently resulted from the disruption of the surface layer of the highly pure monocrystal upon the application of pressure. Measurement of the amplitudes of the frequencies $F(h_6)$ and $F(h_7)$ was impeded by the presence of beat frequencies which experienced strong shifts under pressure.

3. DISCUSSION OF THE EXPERIMENTAL RESULTS ON THE EFFECTS OF PRESSURE ON THE FERMI SURFACES OF FERROMAGNETIC TRANSITION METALS

As was noted above, in the Stoner band model of ferromagnetism, the electron energy levels are split into majority and minority spin subbands by the exchange interaction, so that collectivization of the electrons results in a band energy gap between the majority spin subband and the minority spin subband. The spontaneous magnetic moment is thus due to the unequal number of electrons in the spin subbands. Rigorously speaking, the energy gap, or exchange splitting (Δ), is not a constant, but a strong function of the position of the wave vector k in the Brillouin zone. Even in the hard-zone model, i. e., $\Delta = $ const, the Fermi surfaces for electrons with opposite spin may have quite different topologies because of the shift in the energy ϵ_F relative to the base of the conduction band. In nickel, for example, the Fermi surface in the majority spin subband resembles that of copper, while the minority spin subband Fermi surface resembles that of palladium.

It might be well to point out the special role played by pressure in the band model of magnetism. The parameters of the ground state change under isotropic compression; this will enable us to determine the applicability of various theoretical ideas to describe the system under various external perturbations. Several effects requiring special explanation—for example, two ferromagnets (iron and nickel) having identical pressure derivatives for the magnetization and compressibility, while the pressure derivatives for individual sections of the Fermi surface differ by nearly two orders of magnitude—occur under compression.

These differences can naturally be regarded as resulting from the specific properties of the band structures in iron and nickel and the characteristics of the variations in band structure with pressure.

In order to understand the observed effects of pressure on the Fermi surface, we must first specify a definite model. If we assume that the dispersion laws for the small sections of the Fermi surface that we have studied can be approximated by quadratics, while the pressure effects can be treated as a homogeneous, isotropic effect on the band structure, we can assume, following Lonzarich and Gold [39], that there are two main contributions to the variation in the small extremal cross-sectional areas of the Fermi surface:

1. The first, the "potential contribution," is due to the change in the crystal potential with decreasing volume (which leads to a change in the widths of the bands) as well as the scale effect resulting from the increase in the size of the Brillouin zone upon compression. Changes of this type in the cross sections of the Fermi surface are typical of nonferromagnetic transition metals, and were discussed in detail in the review by Svechkarev and Panfilov [40].

An approximate estimate for this contribution can be obtained from data on the Fermi surface cross section as a function of pressure for non-magnetically ordered metals having the same crystalline structure and an electronic spectrum near the Fermi level similar to that of one of the spin subbands, as well as a compressibility similar to that of the ferromagnet.

Svechkarev and Panfilov's review indicates that the pressure derivatives of the Fermi surface cross sections in groups of nonferromagnetic metals with these properties are of the same sign and similar in size. Metals with these characteristics can thus be used for a qualitative estimate of the potential contribution, and their pressure derivatives can be assumed to be the equal to the potential part of the pressure derivative for the ferromagnet. For example, copper is an obvious model metal for the electron spin subband in nickel.

2. The second contribution to the change in cross-sectional area of the Fermi surface, the "magnetic contribution," is that due to redistribution of electrons between the spin subbands under pressure. The size of this contribution can be estimated within the framework of the Stoner theory. The decrease in magnetization of the ferromagnet with increasing pressure indicates that the exchange splitting decreases with pressure, while the different spin subbands become closer to one another. This results in a redistribution of electrons between the spin subbands and a change in the volumes occupied by the various pieces of the Fermi surface. The following relationship between the change in area of the extremal cross section of the Fermi surface, δS, and the change in the exchange splitting $\delta \Delta$ was suggested by Lonzarich and Gold [39]:

$$\left(\frac{d \ln S_i}{d \ln p}\right)_{\text{magn}} = \pm \frac{m_i^* \Delta_0^i}{e \hbar F_i} \left(\frac{N_\sigma}{N_\uparrow + N_\downarrow}\right) \frac{d \ln M}{d \ln p}, \tag{15}$$

where m_i^* is the corresponding cyclotron mass.

Since all of the quantities on the right-hand side of this equation are known, it should be possible to obtain a numerical estimate of the "magnetic contribution" within the framework of the hard-zone model. However, estimates of this kind are not really appropriate: as was already mentioned above, the subbands with lower and higher populations have somewhat different dispersion laws.

Obviously, when only the "magnetic contribution" is taken into account, the electron Fermi surfaces in the majority spin subband should decrease in size under compression, while the hole Fermi surfaces should increase in size. One might expect the reverse effect in the spin subbands with smaller electron populations: i. e., for the electron Fermi surfaces to increase in size under pressure, while the hole surfaces decrease in size.

The experimental data indicate that the measured pressure derivatives for the lens cross sections in the minority spin subband of iron are quite large and positive. Thus, the main contribution to the pressure derivative is that due to the overflow of electrons from the majority spin subbands into the minority spin subbands. The potential contribution should be of the same order of magnitude as in molybdenum, and should also be positive.

In the case of nickel, however, all of the pressure derivatives are small, and—with the exception of one cross section and field orientation (the X_5 hole pocket in the minority spin subband, with $\mathbf{H}\|[100]$)—do not follow the expected trends for the changes in the cross sections of the various pieces of the Fermi surface under pressure. The observed changes in sign can be regarded as being due to the fact that the potential contribution is larger in absolute value than the magnetic contribution.

Indeed, the pressure derivatives of copper and molybdenum are positive, and equal to 2.5 and 3.9, respectively. The positive pressure derivatives for the electron neck in the majority spin subband and the hole pocket in the minority spin subband ($\mathbf{H}\|[111]$ and $\mathbf{H}\|[112]$) can thus be thought of as indicating that the "potential contribution" dominates in this case.

In cobalt, the pressure derivative for cross section β is negative, in accordance with the fact that "magnetic overflow" has the predominant impact on the variation of this piece of the Fermi surface.

Several other characteristics of the pressure derivative in nickel and cobalt should also be noted. It has been experimentally established that the pressure derivatives for nickel are highly anisotropic; these variations in the pressure derivatives with direction are especially pronounced in the measurements for the X_5 hole pocket. It is known from band structure calculations that the d band comes out to the Fermi level in the neighborhood of this point. Hodges, et al. [41] have observed that the spin-orbit interaction leads to changes in the shape of the hole pocket with magnetic field direction, since the spin-orbit interaction lifts the degeneracy in the energy bands and leads to substantial changes in the energy spectrum.

Another interesting problem is the large value of the pressure derivative for the α-ball in the minority spin subband of cobalt near the point L relative to the pressure derivative for the X_5 hole pocket in the minority spin subband of nickel.

Band structure calculations indicate that the ratio of the density of states at the Fermi energy is approximately the same in the spin subbands of cobalt and nickel. In nickel, $N(\epsilon_F^\downarrow) = 22.3$ and $N(\epsilon_F^\uparrow) = 2.8$; in cobalt, $N(\epsilon_F^\downarrow) = 16.50$ and $N(\epsilon_F^\uparrow) = 6.64$ states/(Ryd · atom · spin), i. e., like nickel, cobalt has a higher density of states at ϵ_F in the minority spin subband than in the majority spin subband.

Thus, the shifts in the Fermi levels resulting from the magnetic overflow of electrons from the majority spin subband into the minority spin subband take place at different rates; the results in [42] indicate that $\delta\epsilon_F^\uparrow = -0.003$ Ryd and $\delta\epsilon_F^\downarrow = 0.0005$ Ryd. Thus, the Fermi level in the majority spin subband should be shift by a substantial amount, while the Fermi level in the minority spin subband should shift very little. This implies that the pressure derivatives for the minority spin subband should be small, and that those for the majority spin subband should be large.

The energy bands experience pressure broadening due to the change in the crystal potential upon compression. The broadening of the bands in the minority spin subband should lead to an increase in the size of the Fermi surface cross sections. On the other hand, the overflow of electrons from the majority spin subband into the minority spin subband associated with the decrease in magnetic moment leads to a decrease in the cross sections of the hole pockets in the minority spin subband and an increase in the cross sections of the electron pockets.

Taking the contribution to the pressure derivative for the X_5 hole pocket in the minority spin subband of nickel resulting from this change in potential into account yields a resultant value of nearly zero for the pressure derivative. A similar situation should apparently also be observed in cobalt for the "normal" hole orbits (as opposed to hybrid hole orbits like α) in the minority spin subband, especially the ϵ and η hole orbits centered on the point L (the dumbbell and ball cross sections, respectively), as confirmed by the calculations in [25].

However, the electron surfaces in the majority spin subband should have large pressure derivatives. This result is supported by the experiments of Anderson, et al. [35], who measured the effect of pressure on the DHVA frequency of the neck in the hyperboloid electron surface of cobalt near the point Γ in the majority spin subband (DHVA frequency β). Their value was $d\ln S/dp = (-2.2 \pm 0.3) \cdot 10^{-3}$ kbar^{-1}.

In the majority electron spin subband, both experiment and theory yield large positive values for the pressure derivative of the electron β orbit: $PD_{\text{theor}} = 16 \cdot 10^{-4}$ and $PD_{\text{exp}} = 22 \cdot 10^{-4}$ kbar^{-1}. This is consistent with the larger shift in the Fermi level of the majority spin subband as a result of electron overflow. A negative sign would indicate that the overflow effect becomes predominant with decreasing volume; broadening of the bands always yields a positive effect.

The large, negative pressure derivative for the α hole orbit in the minority spin subband of cobalt can be explained by the fact that it is a hybrid orbit formed by the intersection between the hole ellipsoid centered on L and a protuberance from the hole monster. The dimensions of this hole orbit will naturally change substantially under compression due to the simultaneous changes in the other two pieces of the Fermi surface.

It should be noted that two quite interesting facts follow from the band structure calculations. First, the density of states at ϵ_F in the minority spin subband and the majority spin subband experience large changes with the difference between them becoming less sharp and the total density of states decreasing under pressure.

Secondly, the calculated values of the size of the exchange splitting and density of states imply that the Stoner–Wohlfarth band magnetism criterion $(IN(\epsilon_F^{\uparrow,\downarrow}) > 1)$ becomes nearly impossible to satisfy under high compression (5%).

In iron, the ratio of the density of states in the majority and minority spin subband Fermi levels at normal pressure is reversed relative to that in nickel and cobalt. The calculations in [42] indicated that $N(\epsilon_F^{\uparrow}) = 7.0$ and $N(\epsilon_F^{\downarrow}) = 2.2$ states/(Ryd· atom · spin) in iron, with $N^{\uparrow}/(N^{\uparrow} + N^{\downarrow}) = 0.76$ and $N^{\downarrow}/(N^{\uparrow} + N^{\downarrow}) = 0.24$. This means that when the electrons flow from the majority spin subband to the minority spin subband, the majority electron spin subband Fermi level will shift more slowly than the minority electron spin subband Fermi level. Indeed, the estimates in [42] indicate that $\delta\epsilon_F^{\uparrow} = -0.002$ Ryd and $\delta\epsilon_F^{\downarrow} = 0.005$ Ryd for $p = 173$ kbar (a decrease of 3% in the lattice constant). This explains the large positive pressure derivative for the minority spin subband electron lens in iron: it is due to the magnetic overflow. The contribution to the pressure derivative from the variation in potential with pressure is also positive, and can only lead to increases in the pressure derivative.

On the basis of the idea that the magnetic and potential contributions are additive, we can use the measured pressure derivatives for the Fermi surface cross sections in rhenium to verify whether, for example, the ϵ dumbbell cross section belongs to a certain spin subband or not, as well as determine the nature of this piece of the Fermi surface in cobalt. The theoretical pressure results for the magnetic component of the pressure derivative of ϵ dumbbell and the total pressure derivative obtained by adding together PD_{mag} and PD_{pot} are given in Table 3. Note that PD_{pot} is the experimental pressure derivative for the dumbbell cross section in rhenium, multiplied by a coefficient which takes into account the difference in compressibility between cobalt and rhenium.

Table 3 obviously indicates that the ϵ dumbbell in cobalt is hole-like and located in the minority electron spin subband.

The values of the density of states at the Fermi level needed to calculate PD_{mag} using equation (25) were taken from [25].

Table 3. Identification of the ϵ Cross Section of the Fermi Surface in Ferromagnetic Cobalt

Type of Carriers in Spin Subband	PD_{mag}, 10^{-3} kbar	PD_{tot}, 10^{-3} kbar	$PD_{exp}(\epsilon)$, 10^{-3} kbar [39]	Type of Carriers in Spin Subband	PD_{mag}, 10^{-3} kbar	PD_{tot}, 10^{-3} kbar	$PD_{exp}(\epsilon)$, 10^{-3} kbar [39]
Majority Spin Subband, e	-1.31	-0.44	—	Minority Spin Subband, e	0.60	1.46	—
Majority Spin Subband, h	1.31	2.16	—	Minority Spin Subband, h	-0.60	0.26	0.4 ± 0.2

Examination of the theoretical results and comparing the experimental and theoretical data on the effect of pressure on small pieces of the Fermi surfaces in ferromagnetic cobalt, nickel, and iron indicate that the changes in the topology of the Fermi sheet can be understood in terms of the broadening of the bands under pressure and the "magnetic overflow" of electrons from the majority to the minority electron spin subbands which results from the changes in the subbands at the Fermi energy. No fundamental qualitative changes occur with pressure.

4. MAGNETIC STRUCTURE AND FERMI SURFACE OF CHROMIUM

Chromium is an antiferromagnet with $T_N = 311$ K. The antiferromagnetic structure of chromium can be represented by two sublattices, one of which contains the atoms at the lattice points, and the other of which contains those at the centers of the elementary cells. However, the spontaneous magnetic moments at the center and vertices of the cube turn out to be different in size and oriented antiparallel to one another, so that we have a long-wavelength sinusoidal modulation of the magnetic moments which is not commensurate with the lattice period. The wave vector for the spin density wave (SDW) $\mathbf{Q} = \frac{2\pi}{a}(1 \pm \delta, 0, 0)$, where $\delta \simeq 0.05$ at liquid-helium temperatures.

Below the Néel temperature (and all the way down to $T = 122$ K) the magnetic structure of chromium consists of transverse SDWs of amplitude $0.59\mu_B$ whose wave vectors are directed along the edges of the cube; the period of the SDW is different from that of the crystalline lattice: the SDW has a period of approximately $20a$ (where a is the lattice parameter). At lower temperatures ($T < 122$ K), the magnetic structure is different: the SDWs become longitudinal, while remaining noncommensurate with the crystalline lattice.

Current ideas hold that the complex magnetic structure of chromium is due to the characteristics of its Fermi surface in the paramagnetic state—in particular, to the fact that the electron and hole pieces of the Fermi surface (Fig. 8), which satisfy the condition $\epsilon_1(\mathbf{k}) = -\epsilon_2(\mathbf{k} + \mathbf{Q})$, are "nested" [44].

Lomer's work [44] implies that the antiferromagnetic ordering is due to direct exchange interaction between the d-like electron (centered on the point Γ) and hole (centered on the point H) pieces of the Fermi surface, which overlap upon translation by the geometric vector $\tilde{Q}$. This electron–hole pairing stabilizes the SDW in the AFM state. Thus, when the condition $\epsilon_1(\mathbf{k}) = -\epsilon_2(\mathbf{k} + n\mathbf{Q})$ is satisfied for $\epsilon(\mathbf{k}) = \epsilon_F$, the interaction at two points in reciprocal space whose wave vectors differ by $n\mathbf{Q}$ leads to a change in the corresponding piece of the Fermi surface during the transition to the AFM state.

If this condition is satisfied at all points in the electron jack around Γ and the hole octahedron around H, these pieces of the Fermi surface should "annihilate" during the transition to the AFM state, and only the electron pockets and hole ellipsoids at the points X and N, respectively, will remain, so that the electronic structure will be commensurate with the lattice period, and $\delta = 0$. However, this condition is not completely satisfied in paramagnetic chromium: the jack is much smaller in size than the octahedron, which, upon translation by $\tilde{Q} = \frac{2\pi}{a}(1 \pm \tilde{\delta})$,

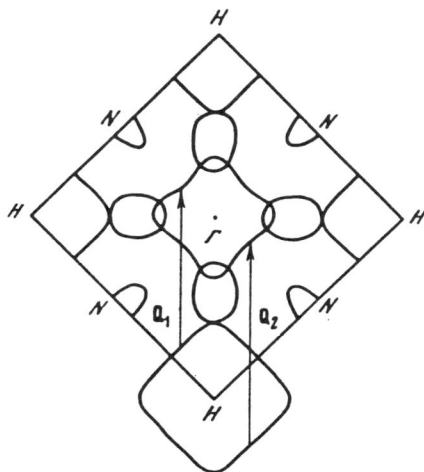

Figure 8   Fermi surface of paramagnetic chromium. The congruency vectors $\mathbf{Q}_1 = \frac{2\pi}{a}(0,0,$ $1 - \delta)$ and $\mathbf{Q}_2 = \frac{2\pi}{a}(0,0,1 + \delta)$, $\delta \approx 0.05$ are indicated by arrows.

leads to a SDW having a period noncommensurate with that of the lattice.

The theoretical calculations in [45, 46] indicate that the parameter $\tilde{\delta}$ (which is determined by the geometry of the Fermi surface) is of order 0.05. This value is in good agreement with neutron diffraction results ($\delta = 0.048 - -0.049$). This good agreement between experiment and the band structure calculations gives us reason to believe that there is a direct relationship between the geometric translation vector $\tilde{Q}$ and the SDW wave vector $\mathbf{Q} = \frac{2\pi}{a}(1 \pm \delta)$ in chromium.

The self-consistent calculations carried out for paramagnetic chromium in the normal and compressed ($p = 270$ kbar) states in [46] showed that pressure has a substantial effect on the electronic structure: the s, p, and d bands are broadened; the the filled portion of the d band also increases in width, while the total width of the filled portion of the conduction band decreases; this can be explained by the fact that some of the s and p electrons flow over into the d band, which has a higher density of states. The base of the d band approaches the base of the conduction band under compression.

In contrast with the non-self-consistent calculations of Lowkes, Kulikov [46] found d-type ellipsoidal hole pockets (which increase in size with pressure) around the N points of the Brillouin zone. The pressure coefficient calculated for the paramagnetic ellipsoid along the [100] axis turned out to be $9 \cdot 10^{-4}$ kbar^{-1}.

Further calculations [47] led to the result that the ellipsoidal pockets are replaced in paramagnetic chromium by nearly spherical hole pockets formed by a level with $N_1'p$ symmetry. The symmetry and volumes occupied by some of the pieces of the Fermi surface change during the transition from the paramagnetic to the antiferromagnetic state. For example, the spherical hole pockets around the N points of the Brillouin zone become ellipsoidal in antiferromagnetic chromium.

The self-consistent calculation for antiferromagnetic chromium indicates that the magnetic moment per atom decreases under pressure. This may possibly be due to electron overflow within the reservoir. This result follows from the fact that the hole pocket experimentally [48] appears to decrease in size under pressure.

4.2. FERMI SURFACE OF CHROMIUM

The Fermi surface model of Graebner and Marcus [49] was used to identify the various cross sections of pieces of the Fermi surface.

In [49], the Fermi surface of antiferromagnetic chromium (which consisted of a chain of hole ellipsoids centered on the N points of the Brillouin zone separated by line segments equal to $n(2\pi/a)\delta$ and chains of electron balls centered on the points X) was obtained by translating the Fermi surface in the paramagnetic phase by the vector $n\mathbf{Q}(n = 1, 2, \ldots)$.

As was shown in [49] and by Fawcett, et al. [50], most of the observed DVHA frequency branches are due to the extremal cross sections of the hole ellipsoid chains shown in Fig. 9. There are four types of extremal cross sections (collective magnetic breakdown orbits) [48]:

1. Extremal cross sections of full ellipsoids with semimajor axes equal to 0.173, 0.234, and 0.268 Å along NH, $N\Gamma$, and NP, respectively (λ- or λ'-orbits).

2. Cross sections originating from the intersection of equivalent elliptical cross sections of the full ellipsoid. Depending on the distance between the centers of the ellipsoids $(n(2\pi/a)\delta)$, these intersections can be of either first ($n = 1$, θ, θ'-, and ν-orbits), second ($n = 2$, ς- and η-orbits), or third ($n = 3$, ϵ-orbit) order. Orbits corresponding to fourth-order intersections have not yet been observed. No higher-order orbits can exist, since the distance between the appropriate ellipsoids would be greater than their maximum diameter.

3. Connected cross sections due to the reunion of two equivalent full-ellipsoid cross sections: first-order π-orbits and second-order ω orbits.

4. Orbits due to the intersection or reunion of two non-equivalent elliptical full-ellipsoid cross sections.

4.3. EXPERIMENTAL RESULTS

We carried out the DHVA effect measurements on polydomain samples. At atmospheric pressure, two frequencies ($F_1 = 4.25\,\mathrm{MGs}$ and $F_2 = 12.5 MGs$) were observed in the type-$\langle 100 \rangle$ samples when the magnetic field was oriented parallel to the type-$\langle 100 \rangle$ axis. These frequencies and their amplitudes turned out to be in good agreement not only with the measurements of Watts [51], which were also carried out on polydomain samples, but also Graebner and Marcus' results for single-domain samples [49]. In accordance with [49, 51], these DHVA frequencies were attributed to the ς (4.25 MGs, centered on N_1) and ν (12.5 MGs, centered on N_3) cross sections, and are due to first- and second-order ellipsoid intersections, respectively.

A third frequency θ ($F_3 = 10.2\,\mathrm{MGs}$), which was attributed to an intersection of first-order ellipsoids, was observed for field orientations along the [110] axis (samples of the second type). Determination of the atmospheric pressure frequency and the pressure derivative for the cross sections corresponding to this frequency was made quite difficult at low frequencies because the large number of harmonics resulting from the fact that two doubly-degenerate frequency branches from the θ and θ' cross sections (Fig. 9) intersect near the [110] axis.

However, these difficulties can be overcome in high-pressure measurements. It is well known from the work of Venema [48] that pressure has practically no effect on the amplitude of the oscillations from cross section θ, while the amplitude from cross section θ' becomes a factor of 1.5 smaller. Using this pressure dependence, we were able to identify cross section θ, improve the value measured for it, and then measure the pressure derivative for this cross section from 5 kbar on up. The

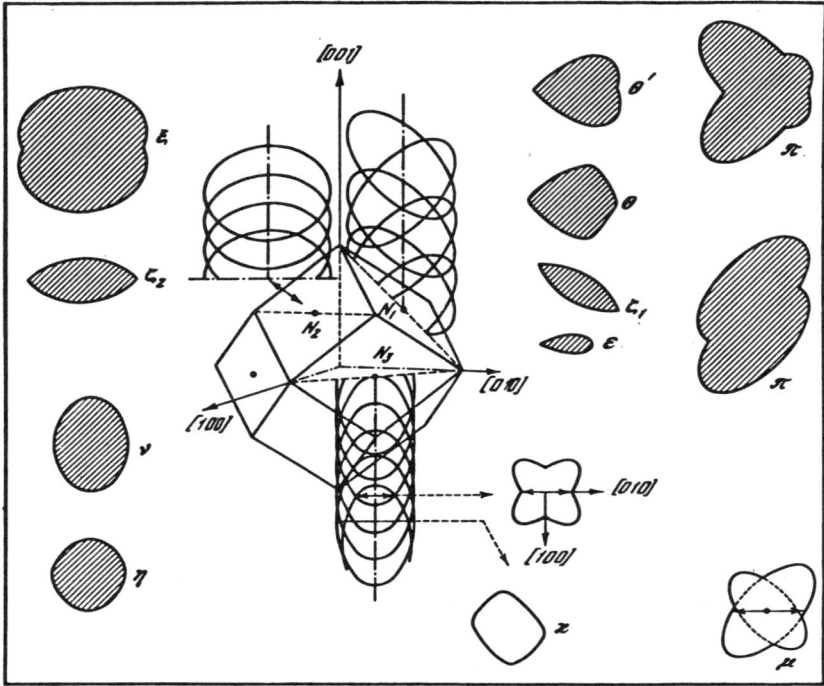

Figure 9   Model of the Fermi surface in antiferromagnetic chromium.

pressure derivatives of the cross sections are given in Table 4 for samples of both types.

As Table 4 indicates, the pressure derivatives for the Fermi surface cross sections are quite large and negative, even though the compressibility of chromium is of the same order of magnitude as that of other magnetically ordered materials $(k_T = -5.25 \cdot 10^{-4}\,\mathrm{kbar}^{-1})$. Thus, significant shrinkage of the extremal cross sections is experimentally observed under compression. According to Fawcett, Griessen, and Vettier [50], this pressure derivative is due to at least two effects: a change in the total cross section of the paramagnetic ellipsoid under pressure, and a change in the incommensurability parameter δ (i. e., "stretching" of the ellipsoids) under pressure. Thus, the extremal cross section can be written as a function of two parameters, S_0 and δ/a, where S_0 is the cross sectional area of the full ellipsoid in the plane containing the orbit of area S.

The experimental data obtained were analyzed using the expressions obtained in [50, 48] for the nth order pressure derivative under the assumption that the ellipsoids undergo isotropic deformation. The experimentally determined pressure derivative can be written as follows:

$$\frac{d\ln S}{dp} = \frac{d\ln S_0}{dp}\left(1 - \frac{1}{2}\frac{d\ln S}{d\ln(\delta/a)}\right) + \frac{d\ln D}{d\ln(\delta/a)}\frac{d\ln(\delta/a)}{dp},$$

where S is the area of the nth-order cross section. The quantity $d\ln S/d\ln(\delta/a)$ is determined by the geometry of the nth-order cross section.

This relation leads to the following relationship between the change in the incommensurability parameter under pressure and the cross section of the Fermi surface for chromium in the PM and AFM states:

$$\frac{d\ln(\delta/a)}{dp} = \frac{\frac{d\ln S}{dp} - \frac{d\ln S_0}{dp}\left(1 - \frac{1}{2}\frac{d\ln S}{d\ln(\delta/a)}\right)}{d\ln S/d\ln(\delta/a)}.$$

The calculated values for the pressure derivative of the incommensurability parameter δ were used to determine the pressure dependence of the SDW wave

Table 4. Pressure Derivatives $(10^{-3}\,\text{kbar}^{-1})$ for Three Cross Sections of the Fermi Surface in Antiferromagnetic Chromium

Number	Extremal cross section		
	ς	θ	ν
1	-13.8 ± 2.2	-3.87 ± 0.37	-2.8 ± 0.16
2	-1.62	-0.51	-0.51
3	0.8	1.4	1.44
4	9.4 ± 1.3	10.6 ± 0.8	9.2 ± 0.3
5	-0.46 ± 0.08	-0.47 ± 0.04	-0.45 ± 0.02

NOTE: (1) values of $d\ln S/dp$ obtained in the present work; (2) $d\ln S/d\ln(\delta/a)$ from [50]; (3) theoretical calculation of $d\ln S_0/dp$ [46]; (4) and (5) are, respectively, $d\ln(\delta/a)/dp$ and $d\ln(Qa/2\pi)/dp$ determined experimentally using the equations in [50].

vector [50]:

$$\frac{d\ln(\mathbf{Q}a/2\pi)}{dp} = -\frac{\delta}{1-\delta}\left(\frac{d\ln(\delta/a)}{dp} + \frac{k_T}{3}\right).$$

The experimentally determined pressure derivatives of the parameter δ (which characterizes the incommensurability of the spin density wave and the lattice parameter) are given in Table 4 along with the pressure derivatives of the SDW wave vector. The table indicates that the experimentally measured pressure derivatives of the incommensurability parameter are strong functions of pressure. This dependence may be due to the fact that the changes in the sizes of the hole octahedron and the jack (and the resulting change in the pressure derivative of the geometric "embedding vector") are not identical. However, the change in the geometric incommensurability parameter is quite small in all of the calculations [46]. The discrepancies between the experimental and theoretical results may be due to the fact that incommensurability effects were neglected in the theoretical calculations.

The values for the pressure derivative of $\mathbf{Q}$ are in good agreement with the results obtained by Fawcett, et al. [52] from direct neutron diffraction measurements, but are much larger than those determined from uniaxial compression of samples or from magnetostriction measurements: $(-0.54 \pm 0.05) \cdot 10^{-3}\,\mathrm{kbar}^{-1}$.

The observed discrepancies in the pressure derivative of the spin density wave vector can be explained in terms of the suggestion of Venema, et al. [48] that small and large changes in relative volume in which the parameter describing the incommensurability between the spin density wave and the lattice parameter is nonzero occur in chromium.

For relatively small changes in volume, the SDW wave vector is "coupled" to the lattice, and the quantity δ is independent of pressure. In this case, $Q \sim 1/a$, and the the SDW wave vector has a small pressure derivative: $\sim 10^{-5}\,\mathrm{kbar}^{-1}$.

For large changes in volume, the SDW becomes "decoupled" from the lattice, so that the parameter δ varies with pressure, and this gives rise to a large total the pressure derivative. The relative changes in volume under uniaxial compression are estimated [48] to be of order 10^{-7}, and can be treated as small. The relative change in volume under isotropic compression is already $\sim 5 \cdot 10^{-3}$ for $p = 10\,\mathrm{kbar}$; this can in no way be considered a small quantity.

The parameter δ was recently [53] found to possess a quality analogous to metallurgical "hardening," i. e., pressure leads to irreversible changes in the SDW wave vector at low temperatures. The equilibrium value of $\mathbf{Q}$ can only be restored by annealing at temperatures greater than $50\,\mathrm{K}$. This effect may obviously lead to slight changes in the pressure derivative when it is determined via magnetostriction measurements, where the sample undergoes cyclic deformations at low temperatures. An explanation of the "hardening effect" in terms of the creation of soliton-like domains due to the interaction between the spin and charge density waves was given by Fenton [54].

REFERENCES

1. Lifshits, I. M., A. M. Kosevich, "The theory of the low-temporary magnetic susceptibility of metals," Zh. Éksp. Teor. Fiz., vol. 29, pp. 730–742, 1955.
2. Kittel, C., "Internal magnetic field in the de Haas–van Alphen effect in iron," Phys. Rev. Lett., vol. 10, pp. 339–340, 1963.
3. Altounian, P., and W. R. Datars, "The DHVA effect in SbTe alloys," Can. J. Phys., vol. 53, pp. 5–9, 1975.
4. Shoenberg, D., and R. Stiles, "De Haas–van Alphen effect in alkali metals," Proc. Roy. Soc. London A, vol. 281, pp. 62–71, 1964.
5. Stark, R. W., and L. R. Windmiller, "Theory and technology for measuring the de Haas–van Alphen type spectra in metals," Cryogenics, vol. 8, pp. 272–281, 1968.
6. Templeton, I. M., "Effect of hydrostatic pressure on the Fermi surface of Cu, Ag, Au," Proc. Roy. Soc. London A, vol. 292, pp. 413–425, 1966.
7. Vol'skii, E. P., A. G. Gapotchenko, E. S. Itskevich, and V. M. Teplinskii, "Effect of pressure on the Fermi surface of molybdenum dioxide and tungsten dioxide," Zh. Éksp. Teor. Fiz., vol. 76, pp. 1970–1974, 1979.
8. Itskevich, I. S., "A high pressure chamber for low-temperature work," Pribory i Tekhnika Éksp., no. 4, pp. 148–151, 1963.
9. Jenings, L. D., and C. A. Swenson, "Effects of pressure on the superconducting transition temperature of Sn, Ta, Te, and Hg," Phys. Rev., vol. 112, pp. 31–43, 1958.
10. Schirber, J. E., "The solid helium pressure generation technique," Cryogenic, vol. 10, pp. 418–422, 1970.
11. Itskevich, E. S., A. M. Sobko, V. A. Sukhoparov, et al., "The effect of pressure on the Fermi surface of cadmium and cadmium–magnesium alloys," High Temp.–High Pressure, vol. 7, pp. 657–658, 1975.
12. Itskevich, E. S., and L. M. Kashirskaya, "Oscillations in the magnetoresistance of electronic indium antimonide near the phase transition under pressure," Fiz. Tverd. Tela, vol. 24, pp. 1129–1132, 1982.
13. Stoner, E. C., "Collective electron ferromagnetism," Proc. Roy. Soc. London A, vol. 165, pp. 372–381, 1938.
14. Anderson, J. R., and A. V. Gold, "De Haas–van Alphen effect and internal field in iron," Phys. Rev. Lett., vol. 10, pp. 227–229, 1963.
15. Baraff, D. R., "De Haas–van Alphen effect and Fermi surface of ferromagnetic iron," Phys. Rev. B–Solid State, vol. 8, pp. 3439–3451, 1973.
16. Gold, A. V., L. Hodges, P. T. Panousis, et al., "De Haas–van Alphen studies of ferromagnetic iron and model for its Fermi surface," Int. J. Magnetism, vol. 2, pp. 357–380, 1971.
17. Tsui, D. C., "De Haas–van Alphen effect and electronic band structure of nickel," Phys. Rev., vol. 164, pp. 670–682, 1967.

18. Tsui, D. C., and R. W. Stark, "De Haas–van Alphen effect in ferromagnetic nickel," Phys. Rev. Lett., vol. 17, pp. 871–875, 1966.

19. Rosenman, I., and F. Batallan, "Low-frequency de Haas–van Alphen effect in cobalt," Phys. Rev. B–Solid State, vol. 5, pp. 1340–1347, 1972.

20. Anderson, J. R., "Electronic structure of ferromagnetic 3d-transition metals at normal and high pressure," In: High Pressure and Low Temperature Physics, eds. C. W. Chu and N. O. Wollan, Plenum Press, New York, pp. 67–86, 1979.

21. Coleman, R. V., R. S. Morris, and D. R. Sellmyer, "Magnetoresistance in iron and cobalt to 150 kOe," Phys. Rev. B–Solid State, vol. 8, pp. 317–332, 1973.

22. Wakoh, S., and J. Yamashita, "Band structure of cobalt by self-consistent procedure," J. Phys. Soc. Japan, vol. 28, pp. 1151–1156, 1970.

23. Ishida, S., "Band structure of HCP cobalt," J. Phys. Soc. Jap., vol. 33, pp. 369–379, 1972.

24. Batallan, F., I. Rosenman, and C. B. Sommers, "Band structure and Fermi surface of hcp ferromagnetic cobalt," Phys. Rev. B–Solid State, vol. 11, pp. 545–557, 1975.

25. Kulikov, N. I., and E. T. Kulatov, "Electronic band structure, Fermi surface, and optical properties of cobalt under pressure," J. Phys. F.: Metal Phys., vol. 12, pp. 2267–2289, 1982.

26. Joseph, A. S., A. C. Thorsen, "De Haas–van Alphen effect and Fermi surface in rhenium," Phys. Rev. A–General Phys., vol. 133, pp. 1546–1552, 1964.

27. Thorsen, A. C., A. S. Joseph, and L. E. Valby, "High field de Haas–van Alphen effect in rhenium," Phys. Rev., vol. 150, pp. 523–529, 1966.

28. Perz, J. M., I. V. Svechkarev, and I. M. Templeton, "Hydrostatic pressure dependence of the Fermi surface of rhenium," Can. J. Phys., vol. 56, pp. 194–199, 1980.

29. Anderson, J. R., F. W. Holroyd, J. M. Perz, et al., "Hydrostatic pressure and uniaxial stress derivatives of the Fermi surface of rhenium," Can. J. Phys., vol. 61, pp. 1428–1433, 1983.

30. Matheiss, L. F., "Band structure and Fermi surface for rhenium," Phys. Rev., vol. 151, pp. 450–454, 1966.

31. Angadi, M. A., E. Fawcett, and M. Rasolt, "High field magnetoresistance and quantum oscillations in iron whiskers," Can. J. Phys., vol. 53, pp. 284–298, 1975.

32. Vinokurova, L. I., A. G. Gapotchenko, and E. S. Itskevich, "Influence of pressure on the de Haas–van Alphen effect and exchange splitting in iron," Pis'ma v Zh. Éksp. Teor. Fiz., vol. 28, pp. 280–283, 1978.

33. Baraff, D. R., "De Haas–van Alphen spectrum and Fermi surface of iron," In: Proc. 13th Internat. Conf. on Low Temp. Phys., NCAR, Boulder, vol. 4, p. 83, 1972.

34. Anderson, J. R., P. Heiman, J. E. Schirber, et al., "Pressure dependence of the electronic structure of nickel," AIP Conf. Proc., vol. 29, pp. 529–530, 1975.

35. Anderson, J. R., J. J. Hudak, and D. R. Stone, "DHVA effect in HCP cobalt," AIP Conf. Proc. vol. 10, pp. 46–50, 1973.

36. Gschneider, K. A., Jr., Solid State Physics, Academic Press, New York–London, Vol. 16.

37. Anderson, J. R., J. J. Hudak, D. R. Stone, et al., "De Haas–van Alphen effect and the influence of pressure on the Fermi surface for HCP Co," In: Proc. Intern. Conf. Magnetism ICM-73, vol. 3, pp. 344–348, Moscow, 1974.

38. Fisher, E. S., and D. Dever, Trans. Met. Soc. AIME, vol. 239, pp. 48, 1976.

39. Lonzarich, G., and A. V. Gold, "Temperature dependence of the exchange splitting in ferromagnetic metals. 1. Information from the de Haas–van Alphen effect in iron," Can. J. Phys., vol. 52, pp. 694–703, 1974.

40. Svechkarev, I. V., and A. S. Panfilov, "Effect of pressure on the electronic structure of transition d-metals," Phys. Status Solidi B, vol. 63, pp. 11–50, 1974.

41. Hodges, L., D. R. Stone, and A. V. Gold, "Field-induced changes in the band structure and Fermi surface of nickel," Phys. Rev. Lett., vol. 19, pp. 655–659, 1967.

42. Vinokurova, L. I., A. G. Gapotchenko, E. S. Itskevich, et al., "The effects of pressure on the electronic structure of ferromagnetic nickel and iron," Zh. Éksp. Teor. Fiz., vol. 76, pp. 1644–1654, 1979.

43. Anderson, J. R., and J. E. Schirber, "Influence of pressure on de Haas–van Alphen frequencies in cobalt," J. Appl. Phys., vol. 52, pp. 1630–1632, 1981.

44. Lomer, W. N., "Electronic structure of chromium-group metals," Proc. Phys. Sci., vol. 80, pp. 489–496, 1962.

45. Rath, J., and J. Callaway, "Energy bands in paramagnetic chromium," Phys. Rev. B–Solid State, vol. 8, pp. 5398–5403, 1973.

46. N. I. Kulikov, "Electronic structure of paramagnetic chromium and its variation under pressure," Fiz. Nizkhykh Temp., vol. 5, pp. 363–366, 1979.

47. Kulikov, N. I., and E. T. Kulatov, "Self-consistent band structure calculation of chromium: Pressure influence," J. Phys. F: Metal Phys., vol. 12, pp. 2291–2308, 1982.

48. Venema, W. J., R. Griessen, and D. W. Ruesink, "Volume dependence of the Fermi surface and of the spin-density-wave Q-vector in antiferromagnetic chromium," J. Phys. F: Metal Phys., vol. 10, pp. 2841–2856, 1980.

49. Graebner, J. E., and J. A. Marcus, "De Haas–van Alphen effect in antiferromagnetic chromium," Phys. Rev., vol. 175, pp. 669–673, 1968.

50. Fawcett, E., R. Griessen, and D. J. Stanley, "Stress dependence of the Fermi surface of antiferromagnetic chromium," J. Low Temp. Phys., vol. 25, pp. 771–796, 1976.

51. Watts, B. R., "The de Haas–van Alphen effect in chromium and the magnetic structure," Phys. Lett., vol. 10, pp. 275–276, 1964.

52. Fawcett, E., R. Greissen, and C. Vettier, "Volume dependence of the wave vector Q of the SDW," Phys. Transition Metals, no. 39, pp. 592–595, 1978.

53. Ruesink, D. W., J. M. Perz, and I. M Templeton, "Pressure-induced 'hardening' of Q-vector locking in chromium," Phys. Rev. Lett. vol. 45, pp. 734–736, 1980.

54. Fenton, E. W., "Domains in the spin-density-wave phases of chromium," Phys. Rev. Lett., vol. 45, pp. 736–739, 1980.

LOCALIZED MAGNETIC MOMENTS
IN METALS AND ALLOYS WITH
SPIN-DENSITY WAVE INSTABILITY

E.T. Kulatov, N.I. Kulikov, V.V. Tugushev

Abstract This paper concerns substances with band magnetism that are unstable with respect to the formation of spin density waves; spin fluctuations are taken into account here. A theory of antiferromagnetism in the alloys of transition metals with heavy rare-earth metals, in the rare-earth metals themselves, and in alloys between the rare-earth metals was proposed. A theory was also developed for antiferromagnetic ordering in transition-metal compounds in which the Fermi surface has special topology in the paramagnetic phase. This theory was used as a framework for theoretical calculations of the Néel temperature in ordered equiatomic iron–rhodium alloys.

INTRODUCTION

The traditional description of the magnetic properties of solids is based on two alternative approaches. The Heisenberg approach employs the concept of point magnetic moments localized at the lattice sites. The Stoner approach uses a model of interacting electron gas with near-total spatial spin density delocalization. Both cases clearly employ an idealization of the real and rather complex density distribution of the magnetic moment in a crystal:

$$m(\mathbf{r}) = \mu_B[p_\uparrow(\mathbf{r}) - p_\downarrow(\mathbf{r})], \tag{1}$$

where μ_B is the Bohr magneton and $\rho_\sigma(\mathbf{r})$ is the electron density in a given spin subband $\sigma = \uparrow, \downarrow$.

The Heisenberg approach has proven useful for describing the properties of magnetic dielectrics and magnetic impurities in metals and superconductors. On the other hand the Stoner scheme is used for metals with heavily delocalized electron shells of the s-, p-, and occasionally the d-orbitals. The majority of transition metals and their compounds cannot be adequately described within the scope of these idealized concepts, which has resulted in the development of new techniques in the band theory of magnetism [1] that are often called the "self-consistent spin fluctuation theories."

This discussion is entirely applicable to the problem of band antiferromagnetism in metals. However in the majority of cases the antiferromagnetic long-range order is caused by the particular topology of the Fermi surface which produces an instability in the paramagnetic phase and generates a superlattice with a characteristic wave vector $\mathbf{Q}$. The concept of spin density waves [2] has made it possible to qualitatively explain main properties of such metals [3]. The incommensurate sinusoidal antiferromagnetic structure of chromium has been investigated most extensively, although a wide range of compounds may be mentioned (that granted have been investigated to a lesser extent) where there clearly exists instability with respect to the formation of spin density waves [3]. The critical and still unresolved problem is accounting for spin fluctuations for electrons that generate the spin density waves. The noticeable quantitative discrepancies between theory and experiment may be primarily attributed to this fact.

The present study is devoted to another class of band magnetics having spin density wave instability for which it is very important to account for spin fluctuations. Specifically, we are speaking of metals with near-congruent Fermi surface sections in which the total portion of such sections within the total Fermi surface is relatively small. Moreover we will assume that the system contains localized (in the sense of expression (1) magnetic moments. We will consider two models related to different origins of the localized moments.

1. The moments are formulated at the internal $4f$ shells of the rare-earth ions in the transition metal matrix (or in the field of the valence s–d-electrons of the rare-earth element itself). It is convenient to describe such moments within the scope of the local spin approximation. Sections 2–4 are devoted to a discussion of the theory of antiferromagnetism in such systems (dilute Cr alloys with Yb, Er; yttrium alloys with heavy rare-earth metals, as well as the rare-earth metals themselves and alloys between rare-earth metals).

2. The moments are formed by the d-electrons in the valence shells of the transition metals and cannot be described within the scope of the local spin model. Sections 5 and 6 discuss the theory of antiferromagnetic ordering in intermetallic alloys and transition metal compounds in which a portion of the Fermi surface has congruent regions and is responsible for the long-range order, while the remaining portion of the Fermi surface generates the local magnetic moments. The alloys FeRh and Pt_3Fe in the greatest detail.

The characteristics of these systems are due to the following circumstances. First, in describing the magnetic ordering of the localized moments, we cannot restrict our discussion to a simple accounting of the indirect exchange via the conduction electrons (the Ruderman- Kittel-Kasuya-Yosida interaction (RKKY)) and neglect the interaction between these electrons. Incorporating the interelectron interaction in systems with spin density wave instability results in a new exchange mechanism for the localized spins via the Bose condensate of triplet electron-hole pairs. There is no basis for neglecting this interaction in comparison to the usual RKKY interaction. Moreover, the presence of localized spins in the system of band electrons having congruent Fermi surface regions initiates spin density waves even in the case of such strong distortion of the congruent regions in the Fermi surface and such a low triplet interaction constant that in the absence of local magnetic moments the system of band electrons ends up far from the antiferromagnetic transition. The localized magnetic moments serve as the source of the order parameter in the self-consistent equation for the spin density wave. The resulting spin density wave in turn orients the localized moments so that their projection at each lattice point corresponds to the amplitude of the resulting spin density wave (Sections 2 and 3).

Second, the contribution of the fluctuations in the spins of the electrons which generate the spin density wave to the total energy of the system is small compared to the contribution from the fluctuations in the local-moment subsystem (due to the significant differential in the density of electron states corresponding to electrons of both types). This means that, to high accuracy, we can limit ourselves to the self-consistent field approximation when describing the spin density waves. At the same time, we require a more refined approach in the spirit of self-consistent spin-fluctuation theory (Section 5) to describe the localized-moment subsystem.

1. $4f$-LOCAL MAGNETIC MOMENTS IN ANTIFERROMAGNETIC METALS. THEORETICAL AND EXPERIMENTAL BAND STRUCTURE RESULTS

Extensive experimental research and band structure calculations have clearly established that the existence of near-congruent electron and hole Fermi surface regions is hardly an exception. The most familiar example of such a system is chromium and its many alloys [3]. However, in their review, Kulikov and Tugu-shev [3] did not consider alloys of chromium with rare-earth metals, i.e., systems in which the impurity atom has a local magnetic moment. Yttrium and scandium alloys with heavy rare-earth metals, as well as the rare-earth metals themselves (with the exception of gadolinium) and alloys between the rare-earth metals [4, p. 64-89] are other examples of such systems. We shall now discuss the experimental results in greater detail.

The heavy rare-earth metals have very low solubility in chromium [5, p. 21-27], so it is very difficult to obtain homogeneous samples. A study of the magnetic susceptibility of $Cr_{1-x}Yb_x$ $(x = (0.7\text{--}11.2) \cdot 10^{-4}\%)$ and $Cr_{1-x}Re_x$ $(x = (0.7\text{--}4.4) \cdot 10^{-4})$ alloys with a high degree of homogeneity as a function of temperature showed that the Curie-Weiss law is satisfied in the paramagnetic domain; this is a direct result of the existence of local magnetic moments. Below the Néel temperature the behavior of the magnetic susceptibility is independent of both the Curie-Weiss paramagnetism and the band antiferromagnetism of pure chromium:

the magnetic susceptibility becomes nearly independent of temperature. This particular behavior obviously may be attributed to the interaction between the $4f$-local magnetic moments of the heavy rare- earth metals and the spin density wave characterizing the band antiferromagnetism of chromium.

Since the spin density wave in Cr is incommensurable with the crystalline lattice and the impurity atoms chaotically displace the chromium atoms, the local magnetic moments of the heavy rare-earth metals are also oriented chaotically, in spite of the fact that this orientation is not random for any lattice site, since it is given by the amplitude of the spin density wave at that point. Thus, the spin density wave "freezes" the local magnetic moment; in this case the local mean value of the moment will be nonzero: $\langle S \rangle \neq 0$, even though the average over the system as a whole is equal to zero. This behavior was called "spin-glass-like" in [6]. However, we cannot treat this as a complete analogy: a neutron-diffraction experiment in this case should reveal the existence of long-range antiferromagnetic order associated with the spin density wave and the local spins frozen in the spin density wave structure.

Such neutron diffraction experiments have also been conducted for some of the systems with spin density waves and local magnetic moments discussed here. Alloys of yttrium with heavy rare-earth metals (Tb, Dy, Ho, Er, Tm) form a continuous series of solid solutions, and antiferromagnetic ordering occurs at all concentrations: either a helicoidal structure with the helicoid axis lying on the c axis of the hexagonal lattice in the metal (for Tb, Dy, and Ho alloys) or a sinusoidal structure modulated along the same axis (for Er and Tm alloys) is formed [4]. A transition to the ferromagnetic state occurs in all concentrated $Y_{1-x}R_x$ alloys with helicoidal order at a T_c less than the Néel temperature T_N. No such transition was observed in the alloys with sinusoidal ordering, although it may exist at very small yttrium concentrations ($x \to 1$).

The yttrium–gadolinium alloy occupies a special position; the helicoidal antiferromagnetic phase exists only to concentrations of order 70% Gd. The ferromagnetic phase begins to become apparent above 60% Gd at low temperatures, and a first order antiferromagnetic-ferromagnetic transition line appears in the T–c phase diagram together with a Lifshits point at concentrations of 60–70%.

Neutron diffraction investigations have revealed that the angle of rotation of the helicoid $\omega = \frac{1}{2}Qc$ (Q is the helicoid wave vector and c is the period of the lattice along the c axis) is a function of concentration and temperature. We emphasize that (unlike Cr) pure Y is paramagnetic at all temperatures, and helicoidal antiferromagnetism is present even for very low concentrations of heavy rare-earth metals. Thus, the $4f$-local impurity moments initiate long-range magnetic order throughout the entire system. The angle of rotation of the helicoid, $\omega \approx 50$, is very similar to the angle of rotation in the pure heavy rare-earth metals near T_N.

The structure of the valence bands of yttrium is similar to that of the bands in heavy rare-earth metals with a hexagonal close-packed crystalline lattice. Moreover, yttrium has the same number of valence electrons $(Z_v = 3)$ as the heavy rare-earth metals. Consequently the Fermi surface geometry of yttrium and the heavy rare-earth metals will be similar. Indeed, the results of band structure calculations for yttrium [7, 8] and various heavy rare-earth metals (see, for example, [4]) show a similarity in the shapes and sizes of the paramagnetic Fermi surfaces. Fig. 1 shows cross-sections of the yttrium Fermi surface along the principal high-symmetry planes of the hexagonal close-packed Brillouin zone. It is clear from the diagram that congruent electron and hole regions exist parallel to the line LH. We may write the following approximate expressions for the bands associated with these regions:

$$\epsilon_1(\mathbf{k}) = \xi(k_z) + \tilde{\mu}_1(\mathbf{k}_\perp), \quad \epsilon_2(\mathbf{k} + \tilde{\mathbf{Q}}) = -\xi(k_z) + \tilde{\mu}_2(\mathbf{k}_\perp), \qquad (2)$$

where $\xi(k_z) = \nu_F k_z$.

The quantity $\mathbf{Q}$ is the geometric overlap vector between the electron and hole pieces, while $\mu_{1,2}(\mathbf{k}_\perp)$ describes the degree of irregularity in the Fermi surface, and the z axis lies on the hexagonal axis (the ΓA direction in the Brillouin zone), while the vector $\mathbf{k}_\perp$ lies in the plane $LHKM$.

Conditions (2) are identical to the conditions on the spectrum determining the special Fermi surface sections of chromium that result in the formation of spin density waves with the vector $\mathbf{Q} \approx \tilde{Q}$ [8]. The subscripts 1 and 2 for the bands are not important in what follows, and will be omitted.

The absence of spin density waves in pure Y is obviously due to the large amount of irregularity in the congruent regions of the Fermi surface and the low triplet interaction constant between the electrons and holes in these regions. However, band structure calculations for Y indicate that the wave vector $\tilde{Q}$ for the Fermi

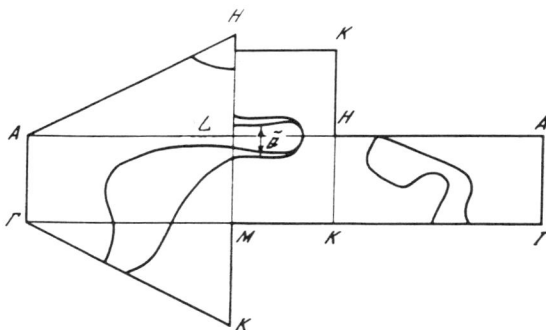

Figure 1   Cross-sections of the Fermi surface in paramagnetic hexagonal yttrium along the principal high-symmetry planes. Q is the congruence vector connecting the plane electron and hole pieces of the Fermi surface.

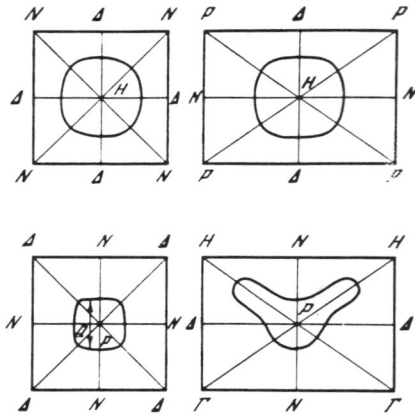

Figure 2   Cross-sections of the Fermi surface in body-centered cubic europium. The electron surface ("superegg") is centered on point H, while the hole surface ("tetracube") is centered on point P.

surface of pure Y is almost identical to the wave vector Q for the helicoidal antiferromagnetic structure in yttrium alloys. It should be stressed that pure hexagonal trivalent rare-earth metals with band structure similar to yttrium (another exception here is gadolinium) have similar pieces of the Fermi surface where the wave vector is likewise very close to the antiferromagnetic structure vector.

As will be demonstrated below, the existence of local magnetic moments leads to the generation of a spin density wave even when there are significant distortions from ideal congruence; the spin- density wave in turn alters the local moments at each site in accordance with the magnitude and sign of the exchange field from the spin density wave. A unique indirect exchange mechanism is implemented via the Bose condensate of the triplet electron-hole pairs [9].

And so, we suggest that interaction between the spin density wave and the $4f$-moments is the primary mechanism responsible for the helicoidal antiferromagnetic structure in all of the heavy rare-earth metals and their associated alloys. However, bivalent europium with a body-centered cubic lattice has a Fermi surface that is completely different from the Fermi surface of yttrium and the heavy rare-earth metals, even though helicoidal magnetic ordering incommensurate with the crystalline lattice exists in this case. The wave vector $Q = 0.29 \cdot 2\pi/over a$. Even in this case, band structure calculations for body centered cubic europium [10-13] show the existence of congruent pieces of the Fermi surface (Fig. 2) where the calculated geometric overlap vector $\tilde{Q} = 0.33 \cdot 2\pi/a$ is in fairly good agreement both in magnitude and direction with experiment [3] (the spin density wave forms along one of the cubic axes). This indicates that the conditions in (2) are also satisfied for this metal and it may be described by the model to be discussed below.

2. MODEL HAMILTONIAN AND SELF-CONSISTENT EQUATIONS

The problem of the behavior of local magnetic moments in a system with spin density wave instability first arose for alloys of Cr with $3d$-magnetic metals [14]. This situation was discussed in the mean-field approximation within the framework of the exciton dielectric model in [9, 15]. All of these studies mentioned above discussed a spin density wave commensurate with the lattice in the limit of small impurity magnetic moment concentrations. This model was used in [16] to describe the helicoidal antiferromagnetic ordering in Y alloys containing heavy rare-earth metals for a broad range of compositions.

The Hamiltonian for the system of band electrons interacting with the local magnetic moments is of the form

$$H = H_0 + H_{ee} + H_{eS} \tag{3}$$

where

$$H_0 = \sum_{k\alpha} \epsilon a_{k\alpha}^+ a_{k\alpha},$$

$$H_{ee} = \sum_{kk'q\alpha\beta} g(k, k', q) a_{k\alpha}^+ a_{k'-q\beta} a_{k+q\alpha},$$

$$H_{eS} = - \sum_{nkk'q\alpha\beta} J_n(k, k') (\hat{S}\hat{\sigma})_{\alpha\beta} a_{k\alpha}^+ a_{k'\beta}, \tag{4}$$

$$J_n(k, k') = J(k, k') \exp\left[i(k - k')R_n\right],$$

$$J(k, k') = \int \psi_k^*(r) J(r) \psi_{k'}(r)\, dr,$$

Where $J(r)$ is the interaction potential between the electron and the localized spin $\hat{S}$; the $\hat{\sigma}$ are the Pauli matrices; $(\alpha\beta)$ are the spin indices; $\mathbf{R}_n$ is the direct lattice vector; (a^+, a) are the electron creation and annihilation operators.

Two types of ordering (antiferromagnetic and ferromagnetic) compete in these alloys and hence we must incorporate two order parameters. The antiferromagnetic order parameter $\Delta_{\pm Q}^{\alpha\beta} = g(Q) \sum_k \langle a_{k\alpha}^+ a_{k\pm Q\beta} \rangle \sigma_{\alpha\beta}$ characterizes the magnetization of the two sublattices. The ferromagnetic order parameter $\lambda^{\alpha\beta} = I(0) \sum_k \langle a_{k\alpha}^+ a_{k\beta} \rangle \sigma_{\alpha\beta}$ characterizes the mean magnetization. The order parameter for the helicoidal structure has symmetry

$$\Delta_Q^{\uparrow\downarrow} = (\Delta_{-Q}^{\downarrow\uparrow})^* \neq 0, \quad \Delta_{-Q}^{\uparrow\downarrow} = \Delta_Q^{\downarrow\uparrow} = 0.$$

while the order parameter for the sinusoidal structure has symmetry

$$\Delta_Q^{\uparrow\downarrow} = \Delta_{-Q}^{\downarrow\uparrow} = \Delta_{-Q}^{\uparrow\downarrow} = \Delta_Q^{\downarrow\uparrow}.$$

The terms containing the z component of the parameter $\lambda^{\alpha\beta}$, i.e., $\lambda^{\alpha\alpha}$ can always be rewritten in terms of the spin-dependent chemical potentials $\mu_\alpha = \mu + \lambda^{\alpha\alpha}$. In our case $\lambda^{\alpha\alpha}$ describes the magnetization along the hexagonal axis, while $\lambda^{\alpha-\alpha}$ describes the magnetization in the basis plane.

The "zeroth-order" Green's functions take the form

where $\tilde{\mu}_\alpha = \mu_\alpha + J_0 \langle\!\langle \hat{S}_n \hat{\sigma} \rangle\!\rangle_{a\alpha}$, and the symbol $\langle\!\langle \ldots \rangle\!\rangle$ denotes averaging over all local moments and the quantum-mechanical mean.

We will first consider the case of a helicoidal spin density wave. The system of equations for the Green's functions may be written as

$$G_{kk}^{\alpha\alpha} = \overset{0}{G}_{kk}^{\alpha\alpha} - \overset{0}{G}_{kk}^{\alpha\alpha} \, \tilde{\mathfrak{J}}_{\pm Q}^{\alpha-\alpha} \, G_{k\pm Qk}^{-\alpha\alpha} - \overset{0}{G}_{kk}^{\alpha\alpha} \tilde{\lambda}^{\alpha-\alpha} G_{kk}^{-\alpha\alpha},$$

$$G_{kk}^{-\alpha\alpha} = -\overset{0}{G}_{kk}^{-\alpha-\alpha} \tilde{\lambda}^{-\alpha\alpha} G_{kk}^{\alpha\alpha},$$

$$\tag{5}$$

$$G_{k\pm Qk}^{-\alpha\alpha} = -\overset{0}{G}_{k\pm Qk\pm Q}^{-\alpha-\alpha} \tilde{\Delta}_{\mp Q}^{-\alpha\alpha} G_{kk}^{\alpha\alpha};$$

$$\Delta_{\pm Q}^{\alpha-\alpha} = g_Q \, i \sum_k G_{k\pm Qk}^{\alpha-\alpha}, \quad \lambda^{\alpha\beta} = I(0) i \sum_k G_{kk}^{\alpha\beta},$$

$$\tilde{\lambda}^{\alpha-\alpha} = \lambda^{\alpha-\alpha} + J_0 \langle\!\langle \hat{S}_n \hat{\sigma} \rangle\!\rangle_{\alpha-\alpha}, \tag{6}$$

$$\tilde{\Delta}_{\pm Q}^{\alpha-\alpha} = \Delta_{\pm Q}^{\alpha-\alpha} + J_{\pm Q} \langle\!\langle \hat{S}_n \hat{\sigma} \exp(\mp iQR_n) \rangle\!\rangle_{\alpha-\alpha}.$$

Note that it is easy to incorporate an external magnetic field in system of equations (5) and (6) by carrying out the substitutions $\tilde{\lambda}^{\alpha-\alpha} \to \tilde{\lambda}^{\alpha-\alpha} + \mu_B (\mathbf{H}\hat{\sigma})_{\alpha-\alpha}$ and $\tilde{\mu}_\alpha \to \tilde{\mu}_\alpha + \mu_B (\mathbf{H}\hat{\sigma})_{\alpha\alpha}$.

Generally speaking, the vector $\mathbf{Q}$ of the spin density wave does not coincide with the geometric overlap vector of the Fermi surfaces $\tilde{\mathbf{Q}}$, and the equations for the sinusoidal structure are then not in agreement with the helicoidal structure. However, we will assume $\mathbf{Q} = \tilde{\mathbf{Q}}$ in order to simplify the calculations in our model, and, by virtue of the symmetry in spectrum (2), all of the results derived above will then be identical for both cases.

Experiment shows that $\mathbf{Q}$ is a strong function of concentration and temperature. We can evidently attribute the concentration dependence to the change in shape of the special Fermi surface regions with concentration, although the temperature dependence of $\mathbf{Q}$ cannot be directly related to the corresponding changes in $\tilde{\mathbf{Q}}$. We know that in systems with spin-wave instability (such as pure Cr), the Q and Q vectors indeed do not coincide [17], but rather are related by a nonlinear equation whose parameters are temperature-dependent. This is not critical for qualitative discussion of the phase diagram and we shall assume the two vectors to be equal. Moreover, we shall for simplicity assume that $\tilde{\mu}_1(\mathbf{k}_\perp) = \tilde{\mu}_2(\mathbf{k}_\perp) = \mu_0 k_\perp / k_f$.

The problem has now been reduced to simultaneous solution of the self-consistent equation for the order parameter Δ_Q and the equation for the average local magnetic moment at the lattice site.

We shall now calculate the average local moment at the nth site $\langle \hat{\mathbf{S}}_n \hat{\sigma} \rangle_{\alpha\beta}$. We may write Hamiltonian (4) averaged over the electron variables in the form of an interaction between the local spin and the effective field H_{eff}:

$$H_{eS} = -\sum_n \left[\frac{1}{2}(\hat{S}_n^+ H_{n,\text{eff}}^- + \hat{S}_n^- H_{n,\text{eff}}^+) + S_n^z H_{n,\text{eff}}^z \right]. \tag{7}$$

$$H_{n,eff}^- = \frac{J_Q}{g_Q} \Delta_{+Q}^{\uparrow\downarrow} \exp(-iQR_n) + \frac{J_Q}{I(0)} \lambda^{\uparrow\uparrow},$$

$$H_{n,eff}^z = \frac{J_Q}{g_Q} \Delta_{-Q}^{\downarrow\uparrow} \exp(iQR_n) + \frac{J_Q}{I(0)} \lambda^{\downarrow\uparrow}, \tag{8}$$

$$H_{n,eff}^z = \frac{J_0}{I(0)} \lambda^z.$$

where $\lambda^z = \lambda^{\uparrow\uparrow} - \lambda^{\downarrow\downarrow}$.

Equation (8) can rewritten in terms of x, y, and z components as

$$H_{n,eff}^x = \frac{J_0}{I(0)} \lambda_\perp + \frac{J_Q}{g_Q} \Delta \cos(QR_n), \quad H_{n,\text{eff}}^y = \frac{J_Q}{g_Q} \Delta \sin(QR_n), \quad H_{n,\text{eff}}^z = \frac{J_0}{I(0)} \lambda^z \tag{9}$$

where we have taken into account the fact that the order parameter $\Delta_Q^{\uparrow\downarrow} = \Delta_{-Q}^{\uparrow\downarrow} = \Delta$ and $\lambda^{\uparrow\downarrow} = \lambda^{\downarrow\uparrow} = \lambda_\perp$ are real.

Now, using Hamiltonian (7) to calculate the statistical sum, we obtain the average

$$\langle \hat{S}_n^i \rangle = \frac{H_{n,\text{eff}}^i}{|H_{n,eff}|} SB_S(|H_{n,eff}|), \tag{10}$$

where $i = x, y, z$; B_S is the Brillouin function; S is the atomic magnetic moment;

$$|H_{n,\text{eff}}| = \left[(H_{n,\text{eff}}^x)^2 + (H_{n,\text{eff}}^y)^2 + H_{n,\text{eff}}^z)^2 \right]^{1/2}.$$

The self-consistent equations for the order parameters Δ, λ^z, and $\lambda_\perp$ can be written in closed form, although it is rather difficult to find their general solution. For the alloys under consideration we shall take advantage of the fact that the helicoidal and ferromagnetic structures do not coexist at any temperatures or concentrations. We may then limit our examination to the particular cases of a helicoidal structure and a purely ferromagnetic structure.

Using (10) we obtain, for the helicoid

$$\langle \hat{S}_n^x \rangle = SB_S\left(\frac{J_Q}{g_Q}\Delta\right)\cos(QR_n), \quad \langle \hat{S}_n^y \rangle = SB_S\left(\frac{J_Q}{g_Q}\Delta\right)\sin(QR_n). \tag{a}$$

The equation for the Néel temperature takes the usual form

$$\Delta = \pi T_N g_Q N(\epsilon_F) \int_0^1 dy \sum_{\omega_m=-\omega_{\max}}^{\omega_{\max}} \tilde{\Delta} \frac{|\tilde{\omega}_m|}{\tilde{\omega}_m^2 + \mu_0 y^2}, \tag{11}$$

$$\tilde{\mathfrak{J}} = 0.5 J_Q x S B_S \left(H_{3,\text{eff}} / T \right) + \Delta. \tag{12}$$

In contrast to [16], we shall now take the impurity scattering effects into account in [11]. We know that systems with spin density wave instability are very sensitive to scattering on impurities, phonons, etc. A method for taking these effects into account was proposed in [18]. In the case of yttrium alloys containing heavy rare earth metals, pure heavy rare-earth metals, and Eu, the primary contribution comes from local spin fluctuation scattering since charge fluctuation scattering is either insignificant or altogether absent here. In any case all of these effects can be expressed formally in (11) in terms of the scattering time τ, i.e.,

$$\tilde{\omega}_m = \omega_m \left(1 + 1/(\tau|\omega_m|) \right) \quad \text{and} \quad \omega_m = \pi T_N (2m + 1).$$

We shall use the coherent potential method [19] to calculate the quantity τ, which is related to the spin fluctuation effect by decomposing the higher-order spin correlators in the Gaussian approximation: $\langle \hat{S}^{2k} \rangle = \langle \hat{S}^2 \rangle^k$, where $k = 1, 2, 3, \dots$. Using this decomposition, we can calculate τ^{-1} in the paramagnetic phase, and this is entirely sufficient for determining the temperature of the antiferromagnetic transition. For simplicity, we shall neglect the fact that the inverse time for electron scattering on spin fluctuations is a function of the order parameter (this correction is of the same order as the ratio of the densities of states of the coherent regions to the total state density for the Fermi surface; this ratio is of order 10^{-1} in real systems). This is why scattering on spin fluctuations need only be taken into account in the self-energy parts of the normal Green's functions.

The potential $J_n(k, k')$ in the Hamiltonian for the RKKY interaction between local spins with the conduction electrons (4) is a random function and is nonzero only at the sites where the yttrium atoms have been displaced by rare-earth metal atoms. In the single-cell approximation, the scattering matrix is of the form

$$\hat{T}_n = [\hat{U}_n - \hat{\Sigma}_n(E)] \, [1 - F_n(E)(\hat{U}_n - \hat{\Sigma}_n(E))]^{-1}, \tag{13}$$

where $\hat{U}_n = J_n(r)[\hat{S}\hat{\sigma}]$ is a random quantity; $\hat{\Sigma}_n(E)$ is the coherent potential; $F_n(E) = \langle 0|G_n|0\rangle$; and $G_n = (E - \epsilon(k) + \Sigma_n(E))^{-1}$.

We should emphasize that the fundamental equation of the coherent potential method has a somewhat irregular form:

$$x\langle \hat{T}_n \rangle + (1 - x) \, \frac{-\Sigma_n(E)}{1 + \Sigma_n(E) F_n(E)} = 0. \tag{14}$$

where the T-matrix is in turn averaged over the spin fluctuation configurations [20] at the sites occupied by the rare-earth metal atoms, i.e.,

$$\hat{T}_n = \frac{1}{F_n(E)} \left[(1 + \hat{\Sigma}_n(E) F_n(E) - \hat{U}_n F_n(E))^{-1} - 1 \right] =$$

$$= \frac{1}{F_n(E)} \left\{ \frac{1}{1 + \hat{\Sigma}_n(E) F_n(E)} \left[\sum_{k=0}^{\infty} U_n^k \left(\frac{F_n(E)}{1 + \hat{\Sigma}_n(E) F_n(E)} \right)^k \right] - 1 \right\}. \tag{15}$$

After using the decomposition given above for the higher-order spin correlators, we obtain the following equation for the coherent potential approximation:

$$\Sigma_n(E) = \frac{x \langle U^2 \rangle F_n(E)}{1 - \left[\dfrac{\langle U^2 \rangle^{1/2} F_n(E)}{1 + \Sigma_n(E) F_n(E)} \right]^2}. \tag{16}$$

where $\langle U^2 \rangle = \frac{1}{3} J_Q^2 S(S+1)$.

We shall now use the following semielliptic approximation for the density of states, where energies are measured in units of the band half-width:

$$N(E) = \begin{cases} \frac{2}{\pi}(1-E^2)^{\frac{1}{2}}, & |E| < 1, \\ 0, & |E| > 1, \end{cases} \tag{17}$$

In this model we may write

$$F_n^0(z) = \langle 0| \overset{0}{G}(z)|0\rangle = 2z - 2(z^2-1)^{\frac{1}{2}}, \tag{18}$$

$$F_n(z) = F_n^0(z - \Sigma),$$

where z is the complex energy.

Substituting (18) into (16), we obtain the following very simple expression for the coherent potential:

$$2xDU^2 = \Sigma(0)\left[1 + \frac{4D^2U^2}{(1 - 2\Sigma(0)D)^2}\right], \tag{19}$$

$$D = \Sigma(0) + (\Sigma(0)^2 + 1)^{\frac{1}{2}}.$$

Note that the coherent potential $\Sigma(0)$ in (19) turns out to be purely imaginary. Since $\text{Im}\Sigma(0) = \tau^{-1}$, we may now solve the self-consistent equations subject to the effect of scattering on fluctuations in the local magnetic moments.

It is convenient to write the free energy of the helicoidal phase as a function of two variables: $\tilde{\Delta}$ as defined in (12), and the new variable $\Sigma = \frac{1}{2} J_Q \langle\langle \tilde{S} \rangle\rangle = J_Q x S B\left(\frac{J_Q}{g_Q}\right)$. Then

$$F_n = \frac{\tilde{\Delta}^2 - \Sigma^2}{g_Q} + F_1(\tilde{\Delta}) + F_2(H(\tilde{\Delta}, \Sigma)), \tag{20}$$

where $H_{\text{eff}} = (J_Q/g_Q)(\tilde{\Delta} - \Sigma)$,

$$F_1(\tilde{\Delta}) = \frac{1}{2} \sum_k \int_{-1}^{1} dx \left[-T \ln\left(1 + \exp\left(\frac{-e_i(k) - \mu_0 x}{T}\right)\right)\right], \tag{21}$$

$$F_2(H_{\text{eff}}) = -xT \ln\left|\frac{\sinh[(2S+1)H_{\text{eff}}/2T]}{\sinh[H_{\text{eff}}/2T]}\right|, \tag{22}$$

$$e_i(k) = \pm(\epsilon^2(k) + \tilde{\Delta}^2)^{\frac{1}{2}}. \tag{b}$$

It is easy to verify that varying (20)–(22) with respect to $\tilde{\Delta}$ and Σ leads to equation (11) for the parameter Δ.

In the case of ferromagnetic ordering, $H'_{\text{eff}} = J_0 \lambda^z / I(0)$, since we may without loss of generality assume $\lambda_\perp = 0$ due to the isotropicity of Hamiltonian (4). The quantity λ^z characterizes the dispersion of the spin subbands in the exchange field and can be written in the following form:

$$\lambda^z = I(0)i\sum_k(G_k^{\uparrow\uparrow} - G_k^{\downarrow\downarrow}) = 2\frac{J_0}{[I(0)N(0)]^{-1} - 1}\langle\langle \hat{S} \rangle\rangle, \tag{c}$$

$$\langle\langle \hat{S} \rangle\rangle = xSB_S(H'_{\text{eff}}), \tag{c}$$

Note that since pure yttrium is not ferromagnetic, the constant $I(0)/N(0) < 1$ (see [8]).

The free energy of the ferromagnetic phase in this model can be written as

$$F_f = F_2(H'_{\text{eff}}) + F_1(0) + \frac{J_0}{I(0)}\lambda^z \langle\langle \hat{S} \rangle\rangle - \frac{N(0)J_0^2}{1 - I(0)N(0)}(\langle\langle \hat{S} \rangle\rangle)^2, \tag{d}$$

or, after writing $\langle\langle \hat{S} \rangle\rangle$ in terms of λ^z:

$$F_f = F_2(H'_{\text{eff}}) + F_1(0) + \frac{1}{4}((I(0)N(0))^{-1} - 1)\frac{(\lambda^z)^2}{I(0)}. \tag{e}$$

The quantity $F_1(0)$ in this case is identical to the energy of the paramagnetic phase and may be dropped, treating the difference $F - F_1(0)$ as the free energy of any of the magnetically ordered phases.

We shall now discuss the structure of the formulae for the free energy in greater detail. The term $F_1(\tilde{\Delta})$ is simply the energy of the band electrons in the field of a spin density wave of fixed amplitude $\tilde{\Delta}$. The term F_2 is the local spin energy in the effective exchange field H_{eff}. Finally, the first term in (20) is the energy of a spin density wave interacting with the average field of localized spins, so that $\tilde{\Delta}^2 - \Sigma^2 = \Delta^2 + 2\Delta\Sigma$.

3. PHASE DIAGRAMS AND TRANSITION TEMPERATURES IN YTTRIUM ALLOYS CONTAINING RARE-EARTH METALS

To determine the temperature of the transition to the helicoidal phase, we retain the first nonvanishing terms in Δ and $\tilde{\Delta}$ in (6). Then, from (11) we obtain

$$\tilde{\Delta} = \Delta(1 + \eta/T_N), \qquad \eta = \frac{1}{6}\frac{J_Q^2}{g_Q} S(S+1)x. \qquad (f)$$

After summation in (11) and integrating over the transverse quasimomentum $k_\perp$ (see equation (2)) we obtain

$$\frac{1}{1 + \eta/T_N} = V\left[\ln\left(\frac{\gamma W}{\pi T_N}\right) - \varphi\left(\frac{\mu_0}{\pi T_N}, \frac{\Sigma(0)}{\pi T_N}\right)\right], \qquad (23a)$$

$$V = g_Q N_n(\epsilon_F), \quad \Pi = \ln\left(\frac{\gamma W}{\pi T_N}\right) - \varphi\left(\frac{\mu_0}{\pi T_N}, \frac{\Sigma(0)}{\pi T_N}\right), \qquad (23b)$$

where the function $\phi(x, y)$ takes the form

$$\varphi(x, y) = \frac{2}{x}\sum_{n=0}^{\infty}\left(\text{arctg}\left[\frac{x}{(2n+1)+y} - \frac{x}{2n+1}\right]\right). \qquad (24)$$

The characteristic value of μ_0 (and, strictly speaking, the interaction parameters J_Q, J_0, $I(0)$, and U) will depend on the alloy composition. Since at present the form of this relation is unknown it is sufficient to assume that they are concentration-independent for a qualitative phase diagram. This assumption is justified in that these alloys are isomorphic and isoelectronic, and hence their wave functions, like the Fermi surfaces, will be similar. We note, however, that the change in μ_0 is nonetheless quite significant and the plane Fermi surface regions virtually vanish in pure Gd.

The temperature of the transition to the ferromagnetic state can easily be obtained from (16) as $\lambda^z \to 0$:

$$T_C = \frac{2}{3}xS(S+1)J_0^2\frac{N(0)}{1 - I(0)N(0)}. \qquad (g)$$

However, this temperature only corresponds to the true Curie temperature in the case $T_C > T_N$. For $T_C < T_N$, the system first becomes antiferromagnetic and then the actual transition to the ferromagnetic phase occurs only when the energies of the magnetic phases become equal. Nonzero homogeneous magnetization then immediately sets in, i.e., a first order phase transition will occur. Thus, the difference between the free energies (13) and (18) must be calculated in order to obtain the transition line.

We will now discuss a method of selecting the characteristic parameters for calculating the phase diagram lines. The Stoner constant $I(0)$ may be calculated from the actual wave functions and the density of Y valence electrons obtained in a self-consistent band calculation. These calculations were carried out for Y using the density functional formalism in [18]. The quantity $N(0)$ was also obtained in the band calculation. The quantity J_0 for the ferromagnetic alloy $Y_{1-x}Gd_x$ can then be determined from the experimental Curie temperature. The bandwidth for which the electron and hole regions overlap and the value of the irregularity parameter were also taken from the Y band calculation [8]. The values of the parameters J_Q and U remained free, and were determined by fitting to the experimental values of the Néel temperature in the alloy $Y_{0.4}Gd_{0.6}$.

Note that in the fit we did not attempt to obtain an absolutely exact or the maximum possible agreement between the calculated and measured critical temperatures. The parameters J_0, J_Q, and U were more estimated than fitted. The values of the parameters used to calculate the phase diagrams, $T_N(x)$ and $T_C(x)$, are given in Table 1. Note that we used virtually the same set of parameters for the $Y_{1-x}Tb_x$ alloys as for the $Y_{1-x}Gd_x$ alloys.

We shall first consider the case $\tau^{-1} = 0$, i.e., we shall neglect the effect of spin fluctuations on the spin density wave amplitude. Then, by using the parameters in Table 1, we obtain the phase diagram of the Y-Gd alloy shown in Fig. 3, where we also compare it to experimental results.

The linear behavior of the Curie temperature as a function of concentration is directly implied by (21) and is completely confirmed by experiment. The Lifshits point is attributable to the nonlinear dependence of the Néel temperature on concentration. It is clear from Fig. 3 that this curve is nearly linear only in the limit of small concentrations, i.e., the rate of increase in T_N with increasing concentration decreases as $\mu_0/\pi T_N \to \infty$. The nonlinearity will be less pronounced if μ_0 also increases with increasing temperature. This is precisely the type of increase in μ_0 with increasing x expected in the $Y_{1-x}Gd_x$ alloys, since the flat pieces of the Fermi surface generally vanish in pure Gd [3]. Thus, the actual curve for $T_N(x)$ will be more nonlinear than that shown in Fig. 3.

Fig. 4 shows the calculated phase diagram for a Y–Tb alloy. The Néel temperature was calculated for the same values of the interaction parameter and the band structure parameter as in the case of the Y–Gd alloy, except that the local magnetic moment was replaced by that corresponding to the Tb atom. In addition, a different value was adopted for the parameter J_0 in the ferromagnetic phase. The theoretical results correctly reproduce the phase diagram for this alloy. It would clearly be possible to achieve better quantitative agreement for T_N by varying the parameters J_Q and U. Note, however, that the theoretical curve reproduces

T, 10^{-4} Ryd

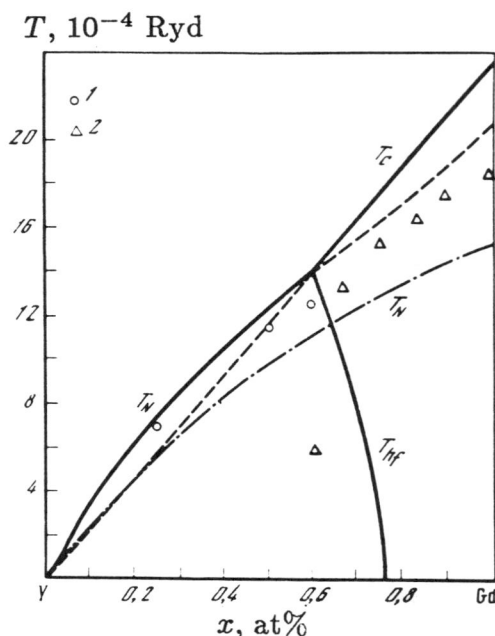

Figure 3   Magnetic phase diagram for Y–Gd alloys Solid lines: Calculated phase transition temperatures $T_N(x)$, $T_C(x)$, and $T_{hf}(x)$ (T_{hf} is the temperature of the the phase transition between the ferromagnetic and helimagnetic states); dashed line: Continuation of the lines $T_N(x)$ and $T_C(x)$ to the antiferromagnetic and ferromagnetic domains, respectively; dot-dash line: Calculation of $T_N(x)$ taking scattering on spin fluctuations into account; 1), 2) Experimental values of $T_N(x)$ and $T_C(x)$, respectively.

the experimentally observed singularity in the function $T_N(x)$ (for the flat piece), although the position of the singularity is shifted in concentration.

The values of the parameters J_Q and J_0 used here are in good (order of magnitude) agreement both with the results of the calculations from "first principles" in [6] and with the estimates in [7].

We shall now take into account the effects of spin density wave scattering on local spin fluctuations. Fig. 3 shows the results of a numerical calculation of $T_N(x)$ for a yttrium-gadolinium alloy using the τ^{-1} obtained from the solution of equation (19) for a given concentration of local magnetic moments are shown. The figure clearly indicates that scattering by spin fluctuations not only leads to a decrease in the Néel temperature, but also makes the function $T_N(x)$ more nonlinear.

Fig. 4 also shows the results of a calculation for $T_N(x)$ in a Y–Tb alloy (taking

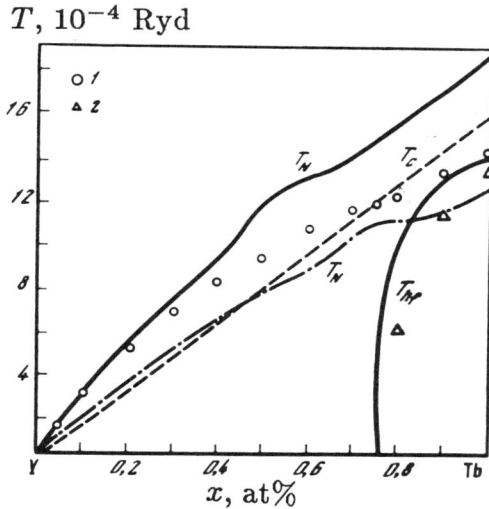

Figure 4 Magnetic phase diagram for Y–Tb alloys Solid lines: Calculation of $T_N(x)$ and $T_{hf}(x)$; dashed curve: Calculation of $T_C(x)$; dot-dash line: Calculation of $T_N(x)$ taking scattering on local spin fluctuations into account; 1), 2) Experimental $T_N(x)$ and $T_C(x)$, respectively.

Table 1. Fundamental Parameters for the Phase Diagram Calculations (All Energies in Rydbergs)

Alloy	s	$N(\epsilon_F)$	$\tilde{Q}$	W	g	μ_0	V	J_Q	J_0
Y–Gd (Fig. 3)	7/2	26.2	0.51	0.05	0.022	0.003	0.024	0.0157	0.0040
Y–Tb (Fig. 4)	6/2	26.2	0.51	0.05	0.022	0.003	0.024	0.0171	0.0017
Y–Gd* (Fig. 3)	7/2	26.2	-	0.05	0.022	0.003	0.025	0.0157	-
Y–Tb* (Fig. 4)	6/2	26.2	-	0.05	0.022	0.003	0.0022	0.017	-

*Taking the effects of scattering from spin fluctuations into account

scattering by spin fluctuations into account), in which the fitting constant V was varied from 0.025 Ry to 0.022 Ry. Good agreement was obtained with experiment, and the singularity in $T_N(x)$ mentioned above (which had previously been shifted in concentration) now coincides with the experimentally observed singularity.

4. THE EFFECT OF A HOMOGENEOUS MAGNETIC FIELD ON THE ANTIFERROMAGNETIC TRANSITION TEMPERATURE

In the model under study, the homogeneous magnetic field **H** may lead to significant changes in both T_N and the nature of the resulting magnetic structure. This effect is the result of three different factors:

1. The magnetic field orients the local magnetic moments and this effect is significant for large values of the ratio $\mu_B H/T_N(x)$, i.e., for large **H** or very low $T_N(x)$.

2. The spin fluctuations are frozen in the magnetic field, i.e., the spin density wave effects discussed above are reduced (this is significant at large $\mu_B H \tau(x)$).

3. Spin polarization of the electrons involved in the formation of the spin density wave occurs in a magnetic field, and this effectively gives rise to a change in the congruence of the electron and hole regions for large $\mu_B H/\mu_0(x)$.

It is, however, obvious that we may neglect the last two effects in the case of yttrium alloys containing heavy rare-earth metals, since $T_N(x) \ll \tau^{-1}$ and μ_0 (see Table 1). Indeed, numerical calculations show that $T_N(x) \ll \Sigma(0)$ at all local moment concentrations, especially since $T_N(x)$ is smaller than the characteristic value of μ_0 used here. If this were not the case, we would be forced to take into account the effect of the magnetic field on the polarization operator defined in (23). However, since the necessary inequalities hold, we may restrict ourselves to only taking the effect of **H** into account in the Brillouin function describing the ordering of the local moments.

We shall first consider the case where the field **H** is parallel to the axis of the helicoid. In this simplest case, we have a conical structure characterized by the presence of magnetization components along the axis of the helicoid. Equation (23) then takes the form

$$\left[1 + \frac{1}{2} xS \frac{J_Q^2}{g_Q T_N} B(y) \right]^{-1} = V \Pi (T_N, \mu_0, \Sigma(0)), \qquad (25)$$

where $y = \mu_B H/T_N$.

At $T > T_N$, all we have is the homogeneous magnetization component induced by the external magnetic field. Below T_N, we have the helicoidal magnetic structure due to the spin density wave and the local magnetic moments aligned with the spin density wave are canted along the field. Note that since the magnetic moment of the spin density wave does not change in magnitude or direction under these conditions, in sufficiently weak fields the change in the projection of the magnetic moment perpendicular to the field with field strength can be used to attempt an experimental determination of the magnetic moment of the spin density wave. The value of the local moment is assumed to be known.

Now let the magnetic field lie in the plane of the helicoid, which we shall assume to lie along the x axis. The situation becomes more complex, and we wind up with two different order parameters, Δ_x and Δ_y, and two equations for the order parameters. The y axis is assumed to be transverse to the magnetic field. When $\Delta_x \to 0$ and $\Delta_y \to 0$, we obtain an equation identical to (25) for the component Δ_y, while the equation for Δ_x takes the form

$$\left[1 + \frac{1}{2} \frac{J_Q^2}{g_Q} \frac{xS}{T_N} \frac{\partial B(y)}{\partial y} \right]^{-1} = V\Pi(T_N, \mu_0, \Sigma(0)). \tag{26}$$

If we let the magnetic field go to zero, we may easily verify that (25) and (26) are in mutual agreement, and agree with (23).

It immediately follows from equations (25) and (26) that the structure with $\Delta_y \neq 0$ forms earlier as the temperature decreases, while Δ_x remains equal to zero. This means that a sinusoidal structure perpendicular to the applied field initially forms in the system. As the temperature decreases, the Δ_x component may in principle become non-zero, i.e., the sinusoidal structure may be replaced by a helicoidal structure. This effect will, however, depend on the magnitude of the external magnetic field and will only exist at moderate magnetic field strengths.

Indeed, consider the limiting case $\mu_B H \gg T_N(x)$, which can be experimentally realized for low heavy rare-earth metal concentrations in Y alloys. It can easily be shown that in this case, (25) leads to the following equation:

$$\left(1 + \frac{1}{2} \frac{J_Q^2}{g_Q} \frac{xS}{\mu_B H} \right)^{-1} = V\Pi(T_N, \mu_0, \Sigma(0)), \tag{27}$$

while (26) yields an expression of the form

$$\left(1 + \frac{1}{2} \frac{J_Q^2}{g_Q} \frac{x}{T_N} \exp\left(-\frac{\mu_B H}{T_N} \right) \right)^{-1} = V\Pi(T_N, \mu_0, \Sigma(0)). \tag{28}$$

It is clear from the last two equations that the structure with $\Delta_y \neq 0$ (equation (27)) comes into existence much earlier than the structure with $\Delta_x \neq 0$. Moreover, equation (28) generally becomes meaningless once the structure with $\Delta_y \neq 0$ comes into existence.

It is interesting to note that in strong magnetic fields, the Néel temperature decreases with increasing field strength (as implied by (27)); the Néel temperature can in principle go to zero at a certain critical value:

$$\mu_B H = \frac{1}{2} \frac{J_Q^2}{g_Q} xS \left(\frac{1}{V\Pi(0, \mu_0, \Sigma(0))} - 1 \right)^{-1}. \tag{29}$$

Thus, the magnetic field reduces the temperature of the transition into the helicoidal antiferromagnetic phase, independent of the direction of the field $\mathbf{H}$ relative to the helicoid vector $\mathbf{Q}$; in the case $\mathbf{H} \perp \mathbf{Q}$, it will not only lead to deformation of the structure in the plane of the helicoid, but lead to the formation of a region where the sinusoidal phase exists. As the field increases, $T_N(x)$ may go to zero, and this will occur most rapidly at low local magnetic moment concentrations x.

This in turn implies that one would expect the appearance of a critical concentration $x_{cr}(H)$ for the transition to antiferromagnetism in the phase diagram for $T_N(x)$ in a magnetic field, with $x_{cr}(H) \sim H$ in accordance with (29).

Of course, our entire discussion here concerns the ordering with wave vector $\mathbf{Q}$ which takes place against the background of the homogeneous magnetization component which is induced by the external field and parallel to it. Note that it is possible to carry out a numerical analysis of equations (27), (28) and the complete self-consistent equations in the magnetic field, taking the effects outlined above (which we have ignored in the present qualitative analysis) into account. However, such a numerical calculation may only be needed to explain experimental measurement results. Here, we shall simply propose methods for carrying out such experiments, and provide qualitative predictions for their results.

Our concluding comment reduces to concluding whether the sinusoidal structure in Y alloys containing Er and Tm can be attributed to relatively small (but anisotropic) additional interactions not included in the present theory. These interactions, which are entirely analogous to the external field, may result in the replacement of the helicoidal structure by sinusoidal ordering.

5. SELF-CONSISTENT SPIN-FLUCTUATION MODEL FOR ANTIFERROMAGNETIC ORDERING IN METALS WITH SPIN-DENSITY WAVE INSTABILITY

In contrast to the case for the rare-earth metals, where the magnetic moments are due to the f-electrons in the internal shells of the magnetic ions, the magnetic moments in transition metals and alloys are due to the band d-electrons (in accordance with equation (1)). Note that the degree of localization of the magnetic moment may differ significantly from the degree of localization of the wave functions, which makes a trivial generalization of the mathematical formalism developed to describe the magnetic properties of substances with localized magnetic electrons impossible (the Heisenberg model, the RKKY model, Anderson superexchange, etc.).

Within the framework of the classical band approach, which takes only single-electron excitations into account (the Stoner model), it is impossible to correctly describe the behavior of d-metals at finite temperatures. This is because collective excitations (spin fluctuations) play the determining role in the temperature dependence of the magnetic properties. Self-consistent spin fluctuation theory (several variants of which were discussed in [1]) is currently the most effective method of calculating the thermodynamic characteristics of band magnetic substances.

These methods are all based on the concept of a magnetic substance as a disordered system in which electrons move in an arbitrary exchange field (the "spin fluctuation field"). Various approximation techniques are used for averaging over the random field configurations: the generalized Hartree-Fock approximation [21, p. 203], the coherent potential method [18, 22], the recursive method [23], the local band method [24], etc.

In the present study, spin fluctuation theory will be used to describe band antiferromagnetism in transition metal compounds in which the Fermi surface has a special topology in the paramagnetic phase. In the majority of the cases under study, the type of antiferromagnetic structure is unambiguously related to these special regions of the Fermi surface, which makes it possible to trace the variations in the structure parameters as a function of temperature, pressure, and composition in a self-consistent manner. The best-studied of these systems from an experimental point of view are the ordered alloys FeRh and Pt_3Fe, as well as a number of other compounds with cubic lattices. These substances were selected since detailed band calculations exist for FeRh and Pt_3Fe; in the absence of such calculations, it is pointless to construct models for specific systems.

A characteristic feature of the systems discussed in this section is the existence of two groups of electron states. On the one hand, this includes the states of the congruent pieces of the Fermi surface which determine the type of antiferromagnetic structure that is present. We note, however, that the contribution from these regions to the total density of states is comparatively small. On the other hand, the majority of the electron states do not directly participate in the formation of the spin density wave and can be described as an electron "reservoir." However, it is essential to note that the interaction between the electrons in the reservoir, which may produce intrinsic spin polarization in the reservoir, cannot be ignored. Exchange interaction between the reservoir and the spin density wave may also have a significant effect on the conditions under which antiferromagnetic long-range order arises in the system.

We shall write the model Hamiltonian in the form

$$H = H_0 + H_r + H_{ext} \tag{30}$$

where

$$H_0 = \sum_{k, \alpha} \epsilon_j(k) a^+_{j k \alpha} a_{j k \alpha} - \sum_{q k \alpha \beta} [(\Delta^q_{12} \hat{\sigma})_{\alpha\beta} a^+_{2 k \beta} a_{1, k-q, \alpha} + \tag{h1}$$

$$+ c.c] + \sum_{q \alpha \beta} \frac{1}{g_t} (\Delta^q_{12} \hat{\sigma})_{\alpha\beta} (\Delta^{-q}_{21} \hat{\sigma})_{\alpha\beta} \tag{h2}$$

is the Hamiltonian of the electrons in congruent regions $j = 1, 2$ written in the mean field approximation [3, 25]; g_t is the triplet coupling constant; Δ^q_{12} is the Fourier component of the spin density wave field; $\hat{\sigma}$ is a vector composed of Pauli matrices; α, β are the spin indices; and c. c. stands for complex conjugate.

We shall use the following expression (see, for example, [26]) for the reservoir Hamiltonian:

$$H_r = \sum_{k \alpha} \epsilon_r(k) c^+_{k\alpha} c_{k\alpha} - \sum_q I_q \hat{m}_q \hat{m}_{-q}; \tag{31}$$

where $\hat{m}_q$ is the magnetic moment operator for the electrons in the reservoir, which can be written in terms of the fermion creation and annihilation operators in the usual way:

$$(\hat{m}_q)_{\alpha\beta} = \sum_k (c^+_k \, \hat{\sigma} c_{k-q})_{\alpha\beta} \tag{32}$$

and

$$H_{ext} = - \sum_{qk\alpha\beta} J_q (\hat{m}_q \hat{\sigma})_{\alpha\beta} a^+_{1k\beta} a_{2,\,k-q,\,\alpha} + \text{c. c.} \tag{i}$$

The quantities I_q and J_q in (31) and (32) are the exchange integrals between electrons in the corresponding bands.

We shall also use the mean-field approximation to describe the interaction between the reservoir electrons and the spin density wave, i.e., we shall everywhere neglect the fluctuations in the spin density wave itself compared to the fluctuations in the reservoir spin density. This approximation is justified, because of the smallness of the ratio of the number of electrons responsible for the spin density wave to the total number of electrons in the spin-polarized reservoir. Hamiltonian (32) may then be written as

$$H_{\text{вн}} = - \sum_q \frac{1}{g_t} J_q \{(\hat{m}_q)_{\beta\alpha} (\Delta^q_{12} \, \hat{\sigma})_{\alpha\beta} + g_t (\langle \hat{m}_q \rangle \hat{\sigma})_{\alpha\beta} \times \tag{33a}$$

$$\times \, a^+_{1k\beta} a_{2,\,k-q,\,\alpha} - \langle \hat{m}_q \rangle \Delta^q_{12} \} + \text{c. c.} \tag{33b}$$

Note that "intraband" terms of the form $(\hat{m}_q \hat{\sigma})_{\alpha\beta} a^+_{jk\beta} a_{j,\,k-q,\,\alpha}$ which contribute to the RKKY exchange interaction were not included in Hamiltonian (30). In contrast to the systems with local magnetic moments, the primary contribution to the interaction between the moments $\mathbf{m}_q$ here comes from direct exchange in the reservoir.

Within the framework of these approximations, the problem naturally breaks down into two stages. On the one hand, we must write a self-consistent equation for the spin density wave amplitude which exactly corresponds to equation (6) derived above for the local-moment case (with $\langle \hat{S}_Q \rangle$ replaced by $\langle \hat{m}_Q \rangle$). On the other hand, the quantity $\langle \hat{m}_Q \rangle$ itself is calculated using spin-fluctuation Hamiltonian (31), taking the effect of the spin density wave magnetizing field (the first term in Hamiltonian (33)) into account.

We shall restrict ourselves to a qualitative discussion of the effect of spin fluctuations on the Néel temperature within the framework of this model, keeping the FeRh and Pt_3Fe systems in mind.

We shall use the representation of Hamiltonian (33) in terms of a continuous integral over random fields $\{\xi_i\}$ in the spirit of the self-consistent spin fluctuation theory approach. Using the Hubbard-Stratonovich transform, we obtain the standard expression for the partition function Z of the reservoir electrons in the spin density wave field:

$$Z = \int Sp \exp \{- \beta [\tilde{H}_0 + \sum_q I_q \xi_q \xi_{-q} - I_q \xi_q m_{-q} - I_q \xi_{-q} m_q]\} \prod_q d\xi_q, \tag{j}$$

$$\tilde{H}_0 = \sum_{k\alpha} \epsilon_r (k) \, c^+_{k\alpha} c_{k\alpha} - \sum_{qk} \frac{1}{g_t} J_q \{(\Delta^q_{12} \, \hat{\sigma})_{\alpha\beta} c^+_{k\alpha} c_{k-q,\,\beta} + \text{c.c.}$$

We shall assume for simplicity that the matrix element I_q is independent of the momentum, so that the spin-fluctuation term of the Hamiltonian may be rewritten in the site representation:

$$H_{sp\text{-}fl} = \sum_i I \{\xi^2 - 2\xi_i \hat{m}_i\}, \quad I \equiv I_q. \tag{k}$$

The simplest calculation method used in spin-fluctuation theories is a statistical approximation for the fields $\{\xi_i\}$ and using the single-site coherent potential technique [18] for averaging. Specifically, our problem involves determining the spatial dependence of the coherent potential:

$$\hat{\Sigma}(\omega, \mathbf{R}_i) = \Sigma_0(\omega)\hat{I} + \Sigma_1(\omega)\cos(\mathbf{Q}, \mathbf{R}_i)\hat{\sigma}, \tag{34}$$

where $\hat{I}$ is the identity matrix; $\mathbf{Q}$ is the wave vector of the spin density waves. Thus, $\Delta_{ij}^q = \Delta^{\pm Q}\delta_{q\pm Q}$. We shall now consider the three-dimensional problem. Thus, the single-site Green's function G for the effective medium should be written in spinor form, as in (34), i.e.,

$$\hat{G}(\omega, \mathbf{R}_i, \mathbf{R}_j) = \overset{0}{G}(\omega)\hat{I} + G(\omega, \mathbf{R}_i)\hat{\sigma}. \tag{35}$$

where

$$\overset{0}{G}(\omega) = \sum_k \frac{\omega - \epsilon_r(\mathbf{k}+\mathbf{Q}) - \Sigma_0(\omega)}{[\omega - \epsilon_r(\mathbf{k}) - \Sigma_0(\omega)][\omega - \epsilon_r(\mathbf{k}+\mathbf{Q}) - \Sigma_0(\omega)] - \tilde{\Sigma}(\omega, \mathbf{R}_i)^2}, \tag{36a}$$

$$G_1(\omega, \mathbf{R}_i) = \sum_k \frac{-\tilde{\Sigma}(\omega, \mathbf{R}_i)}{[\omega - \epsilon_r(\mathbf{k}) - \Sigma_0(\omega)][\omega - \epsilon_r(\mathbf{k}+\mathbf{Q}) - \Sigma_0(\omega)] - \tilde{\Sigma}(\omega, \mathbf{R}_i)^2}, \tag{36b}$$

$$\tilde{\Sigma}(\omega, \mathbf{R}_i) = \Delta_{12}(\mathbf{R}_i) + \Delta_{21}(\mathbf{R}_i) - \Sigma_1(\omega, \mathbf{R}_i) = \tag{36c}$$

$$= [2\Delta_r - \Sigma_1(\omega)]\cos \mathbf{Q}\mathbf{R}_i; \quad \cos \mathbf{Q}\mathbf{R}_i = \pm 1.$$

Note that equations (34)-(36) are only suitable for commensurate antiferromagnetic structures; incommensurate spin density waves will not be considered in the present paper.

We now introduce the random fields

$$\hat{\eta}_i = \Sigma_0\hat{I} - (2I\xi_i - \Sigma_1)\hat{\sigma} = \eta_0\hat{I} + \eta_1^i\hat{\sigma} \tag{l}$$

and write the coherent potential equation for the complete vertex in the single-site approximation:

$$\langle\hat{\Gamma}^i\rangle = \langle\Gamma_0^i\hat{I}\rangle + \langle\Gamma_1^i\hat{\sigma}\rangle = 0, \tag{m}$$

where

$$\Gamma_0^i = [\eta_0 - \overset{0}{G}(\eta_0^2 - \eta_1^{i2})]/\text{Det}(i), \tag{37}$$

$$\Gamma_1^i = [\eta_1^i + G_1^i(\eta_0^2 - \eta_1^{i2})]/\text{Det}(i), \tag{38a}$$

$$\text{Det}(i) = 1 + (\overset{0}{G}{}^2 - G_1^2)(\eta_0^2 - \eta_1^2) - 2\eta_0\overset{0}{G} - 2\eta_1 G_1. \tag{38b}$$

The averaging in (37) is carried out separately over each sublattice in the antiferromagnet.

We thus have an equation for the order parameter (the antiferromagnetic gap Δ) that is formally identical to (11) but requires a different definition of the quantity $\tilde{\Delta}$. The latter is now defined as

$$\tilde{\Delta} = \frac{1}{2}J_Q x\langle m_Q\rangle + \Delta. \tag{39}$$

The mean reservoir band moment $\langle m\rangle$ for the ith sublattice can be determined by averaging over the ensemble of random fields:

$$\langle m_Q^i\rangle = \int e^{-\beta Z(\xi^i)}\xi^i d\xi^i / \int e^{-\beta Z(\xi^i)} d\xi^i. \tag{40}$$

Simultaneous solution of equations (11) and (37)–(40) yields a complete description of the thermodynamics of an antiferromagnetic within the scope of the model proposed here.

6. RESULTS FROM A NUMERICAL CALCULATION OF THE NÉEL TEMPERATURE IN THE SPIN FLUCTUATION THEORY MODEL OF AN ANTIFERROMAGNETIC WITH SPIN-DENSITY WAVE INSTABILITY

In studying the effect of reservoir spin fluctuations on the Néel temperature in equations (37)–(40) we can restrict ourselves to an approximation linear in Δ. A qualitative analysis of the expressions for the Green's function (37) and $\langle m_Q \rangle$ as a function of Δ leads to the following conclusions.

1. In contrast to the local-spin case, we have a noticeable drop in the Néel temperature with identical parameters (g_t, μ) describing the congruent regions of the spectrum. This is due to the fact that the spin orientation effect is suppressed by the spin density wave field due to the spin fluctuations in the reservoir. This effect is absent for fundamental reasons in the case of a system with local spins. We emphasize that we are not discussing the direct effect of the electron scattering responsible for the formation of a spin density wave by the spin fluctuations (which effect was discussed above) on the Néel temperature, but rather a completely different mechanism due to the lifetime of the spin fluctuations themselves.

2. In contrast to the local-spin case, it is not possible to establish long-range antiferromagnetic order in the limit of very low temperatures for any values of the noncongruence parameters. We obtained an opposite result above for the local-spin model. In the present model, there are critical values of the noncongruence parameters where spin density waves cannot form at any temperature. This is formally due to the fact that the constant η in equation (23) for the Néel temperature is now proportional to T_N.

This may easily be shown. Indeed, in the vicinity of T_N, the magnetization $\langle m_Q \rangle$ may be represented within the framework of linear response theory as being directly proportional to the antiferromagnetic gap, i.e.,

$$\left\{ \left(1 + \frac{J_0^2}{g_Q} N_r(\epsilon_F) K(T, <m_Q^2>_0) \right) \right\}^{-1} = g N_N(\epsilon_F) \, \Pi(T), \tag{41}$$

An additional assumption is made in (41): That $\langle m_Q^2 \rangle$ is independent of temperature and is equal to its value at $T = 0$. This assumption is justified by the results of spin-fluctuation theory calculations of $\langle m_Q^2 \rangle$ for most of the transition metals for which the Stoner criterion $N(\epsilon_F) g_{\text{eff}} > 1$ is satisfied (see, for example, [18, 22]). The polarization operator $K(T, \langle m_Q^2 \rangle_0)$ may be calculated as the convolution of the Green's functions $\overset{0}{G}(\omega)$ only, since $G_1(\omega, \mathbf{R}_i)$ is equal to zero in the vicinity of T_N (we are only considering the projection onto the z axis). In this case, equation

(37) reduces to equations (14) and (15) discussed above. However, the meaning of the equation is different here: where before we calculated $\Sigma(E)$, which was determined by electron scattering processes in the exterior system of local spins, $\Sigma(E)$ is now determined by electron scattering processes involving fluctuations in the electron spins themselves.

We now assume the reservoir band to have a semielliptical density of states (17); solving equation (19) then yields the unknown quantity $\Sigma(E)$. Then, in order to calculate $\langle m_Q \rangle$, we must calculate polarization operator $\Pi(T, \langle m_Q^2 \rangle_0)$ in (41) using the Green's functions whose self-energy we have already calculated. We know that the function $K(T, \langle m_Q^2 \rangle)$ can only be calculated analytically for simple models. It is most natural in this case to assume for simplicity that the reservoir can be described qualitatively using the model used in describing the bands involved in

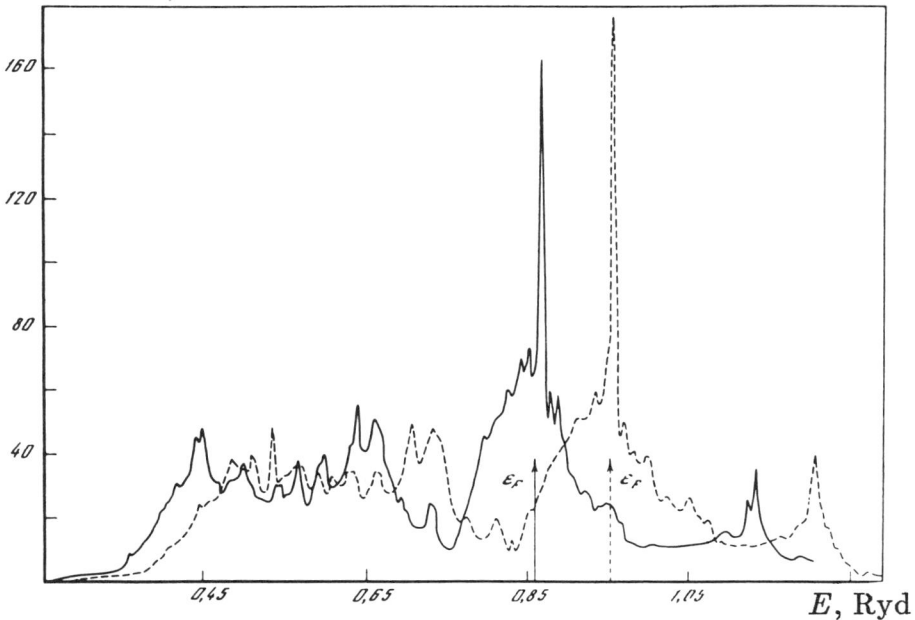

Figure 5   The density of states for the paramagnetic alloy FeRh at atmospheric pressure (solid line) and at 100 kbar (dashed line). The Fermi levels are indicated by arrows.

the formation of the spin density wave, although the noncongruence parameter for the reservoir bands must be selected close to the bandwidth, which will eliminate the spin density wave instability in these reservoir bands. Therefore, under these assumptions, polarization operator K is identical to the operator Π defined in equations (23), (24).

These approximations enabled us to rather easily carry out numerical calculations of T_N as a function of our model parameters. Specifically, note that (23) now takes the form

$$(42)$$

which confirms the qualitative conclusion reached at the beginning of this section that the denominator on the left-hand side of (42) has a finite value as $T \to 0$.

Having in mind a specific antiferromagnetic FeRh alloy with congruent Fermi surface regions and a largely localized magnetic moment density on the iron atoms [29], we carried out numerical calculations of the Néel temperature on the basis of this model. The stoichiometric ordered equiatomic FeRh alloy has a CsCl crystalline lattice. Thus, each elementary cell contains an iron atom, i.e., the concentration of magnetic atoms $x = 1$. The theoretical results in [28] indicate that localized magnetic moment per Fe atom is $3.3\mu_B$. We used the density functional

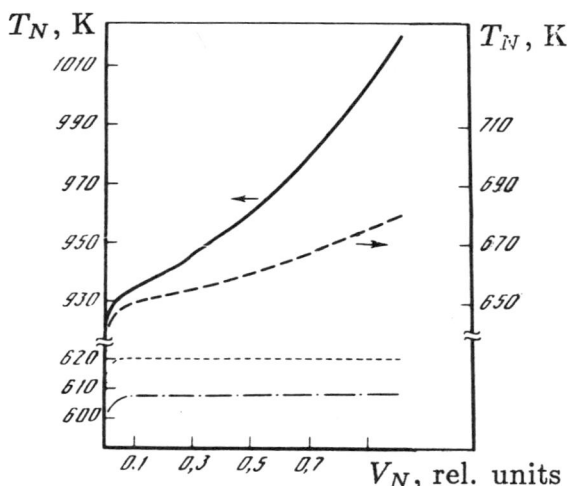

Figure 6   The Néel temperature plotted as a function of the interaction constant V_N for the congruent bands $(V_N + V_R = V)$ Solid line: $V = 4$, $W_N = 0.01$, $\mu_N = 0.005$, $\mu_R = 0.05$; dashed line: $V = 4$, $W_N = 0.007$, $W_R = 0.07$, $\mu_N = 0.002$, $\mu_R = 0.02$; dot–dash line: $V = 9.59$, $W_N = 0.007$, $W_R = 0.07$, $\mu_N = 0.003$, $\mu_R = 0.03$; fine dashed line: $V = 9.59$, $W_N = 0.007$, $W_R = 0.07$, $\mu_N = 0.002$, $\mu_R = 0.02$ (all energies in Ry).

Table 2.   Total and Partial Density of States
$N(\epsilon_F)$ and $N_l(\epsilon_F)$, state/(Ry · elem.cell)
and the Charges $n(\epsilon_F)$ and $n_l(\epsilon_F)$ at the
Fermi Level ϵ_F in a Paramagnetic Alloy

Charac- teristic	$p = 0$		$p = 100\,\mathrm{kbar}$	
	Fe	Rh	Fe	Rh
N_s	0.071	0.052	0.136	0.040
n_s	0.522	0.318	0.465	0.287
N_p	0.256	0.185	0.356	0.256
N_d	51.453	20.048	99.26	21.29
n_d	6.605	5.991	6.639	0.138
N_f	0.277	0.194	0.304	0.239
n_f	0.155	0.073	0.138	
N_{out}	2.094		3.793	
n_{out}	2.154		2.428	
N	74.631		125.684	
n	17.00		16.996	

Notation: N_{out} and n_{out} are the density of states
and charge, respectively, in the interlattice region
(outside the MT-shells).

formalism [29] to calculate the Stoner interaction constant averaged over all bands.
It seems reasonable to assume $g_Q \simeq J_Q \simeq I(\epsilon_F)$. Under normal conditions (at
atmospheric pressure), the dimensionless Stoner constant $V = I(\epsilon_F)N(\epsilon_F) \simeq 4$.
Note that a self-consistent calculation of the band structure under pressure shows
a significant increase in the density of states at the Fermi energy in spite of the
broadening of the bands. This is due to the proximity of ϵ_F to the high, narrow
peak in the density of states, i.e., on the rising portion of the $N(E)$ curve.

Fig. 5 and Table 2 give the results of a calculation of the total and partial density
of states for the alloy FeRh at two values of the lattice constant (corresponding
to atmospheric pressure and $p \approx 100\,\mathrm{kbar}$. Our numerical estimates reveal that
the constant increases to $V = 9.59$ at pressures of order 100 kbar ($\Delta a/a \simeq 2.3\%$).
At these pressures and temperatures of $\sim 600\,\mathrm{K}$, we have a direct second order
transition from the paramagnetic phase to the antiferromagnetic phase.

Note that in accordance with our assumption that $g_Q \simeq J_Q$ we have the equality

$V = V_N + V_R$, where $V_N = g_Q N_N(\epsilon_F)$ and $V_R = J_Q N_r(\epsilon_F)$. Fig. 6 shows the calculated Néel temperatures plotted as a function of the interaction constant. It is clear that the form of this relation and the absolute value of T_N are both largely determined by the width of the congruent and reservoir bands as well as the position of the chemical potential rather than the value of the constant V_N itself.

In conclusion, we can say that the equations for the spin fluctuation theory model in the presence of spin density wave instability derived in the last section may be solved numerically, even though they are more complex than the equations for the local-spin model. In this case, the characteristic model parameters will be determined by a self-consistent calculation of the band structure for each situation (the alloy composition, pressure, and degree of ordering). Only an integrated calculation involving a calculation of the parameters of the ground state $(T = 0)$ from "first principles" and a thermodynamic calculation within the framework of the model described here will enable us to formulate a consistent theory of magnetic ordering in transition metal alloys.

REFERENCES

1. Moriya, T. "Recent developments in the theory of collective electron magnetism," Usp. Fiz. Nauk, vol. 135, pp. 117–170, 1981.
2. Keldysh, L. V., and Yu. V. Kopaev, "A Possible instability of the semimetallic state with respect to the Coulomb interaction," Fiz. Tverd. Tela, vol. 6, pp. 2791–2798, 1964.
3. Kulikov, N. I., and V. V. Tugushev, "Spin density waves and band antiferromagnetism in metals," Usp. Fiz. Nauk, vol. 144, pp. 643–680, 1984.
4. Coqblin, B., The Electronic Structure of Rare-Earth Metals and Alloys: The Magnetic Heavy Rare-Earth, Academic Press, NY, 1977.
5. Elliott, R. P., Constitution of Binary Alloys, McGraw-Hill, New York, 1968, Suppl. 1.
6. Giurgiu, A., I. Pop, M. Popescu, and Z. Gulaesi, "Spin-glass like behavior of dilute Cr–Er and Cr–Yb alloys," Phys. Rev. B–Solid State, vol. 24, pp. 1350–1359, 1981.
7. Loucks, T. L., "Fermi surface and positron annihilation in yttrium," Phys. Rev. B–Solid State, vol. 144, pp. 504–511, 1966.
8. Kulikov, N. I., and E. T. Kulatov, "The electron band structure, Fermi surface and optical properties of hexagonal yttrium," Fizika Nizkikh Temperatur, vol. 9, pp. 823–831, 1983.
9. Kopaev, Yu. V., "A theory of the interrelation between electron and structural transformations and superconductivity," Trudy Fiz. Inst. Akad. Nauk SSSR (Lebedev Phys. Inst.), vol. 86, pp. 3–100, 1975.

10. Andersen, O. K., and T. L. Loucks, "Fermi surface and antiferromagnetism in europium metal," Phys. Rev., vol. 167, pp. 551–556, 1968.

11. Freeman, A. J., and J. O. Dimmock, "Band structure and electronic properties of europium metal," Bull. Am. Phys. Soc., vol. 11, pp. 216, 1966.

12. Fukuchi, M., M. Matsumoto, I. Shibata, et al., "Fermi surface and magnetism of rare earth metal europium for $a = 2/3, 0.8$, and 1," Keio Eng. Rept., vol. 33, pp. 83–95, 1980.

13. Matsumoto, M., M. Fukuchi, Y. Sakizi, and S. I. Kobayashi, "Density of states and isomer shift on the rare-earth metal europium," J. Phys. F: Metal Phys., vol. 13, pp. 1457–1464, 1983.

14. Shibatani, A., "Effect of impurities with localized moments on the spin density wave," J. Phys. Soc. Jap., 1971, vol. 31, pp. 1642–1649.

15. Volkov, B. A., and T. T. Mnatsakanov, "The interaction of localized magnetic moments in systems with electron-hole pairing," Zh. Éksp. Teor. Fiz., vol. 75, pp. 563–576, 1978.

16. Ami, S., and W. Young, "Bulk susceptibilities in itinerant antiferromagnetic alloys" J. Phys. F: Metal Phys., vol. 11, pp. 227–245, 1981.

17. Kulikov, N. I., and V. V. Tugushev, "Magnetic ordering in yttrium alloys with rare earth metals," Fiz. Tverd. Tela, vol. 25, pp. 2442–2448, 1983.

18. Fenton, E. W., "Changes of the Q-vector of the spin density wave in chromium," Solid State Comm., vol. 32, pp. 195–199, 1979.

19. Hasegawa, H., "Single-site spin fluctuation theory of itinerant electron systems with narrow bands," J. Phys. Soc. Jap., vol. 49, pp. 178–188, 1980.

20. Harmon, B. N., and A. J. Freeman, "Augmented-plane-wave calculation of indirect-exchange matrix elements for gadolinium," Phys. Rev. B–Solid State, vol. 10, pp. 4849–4855, 1974.

21. Lindgard, P. A., Magnetism in Metals and Metallic Compounds, Plenum, New York, 1976.

22. Capellman, H., and V. Vieira, "Magnetic properties of ferromagnetic transition metals above T_C: Qualitative aspects," Phys. Rev. B–Solid State, vol. 25, pp. 3333–3349, 1982.

23. Hubbard, J., "The magnetism of iron," Phys. Rev. B–Solid State, vol. 19, pp. 2626–2637, 1979; "Magnetism of iron," Ibid., vol. 20, pp. 4584–4595, 1980.

24. You, M. V., and V. Heine, "Magnetism in transition metals at finite temperatures. 1. Computational model," J. Phys. F: Metal Phys., vol. 12, pp. 177–194, 1982.

25. Korenman, V., J. L. Murray, and R. E. Prange, "Local-band theory of itinerant ferromagnetism," Phys. Rev. B–Solid State, vol. 16, pp. 4032–4062, 1977.

26. Edwards, D. M., "The paramagnetic state of itinerant electron systems with local magnetic moments. 1. Static properties," J. Phys. F: Metal Phys., vol. 12, pp. 1789–1810, 1982.

27. Kulikov, N. I., E. T. Kulatov, L. I. Vinokurova, and M. Pardavi-Horvath, "Electronic band structure and magnetic order in FeRh," J. Phys. F: Metal Phys., vol. 12, pp. L91–L96, 1982.
28. Koenig, C., "Self-consistent band structure of paramagnetic ferromagnetic and antiferromagnetic ordered FeRh," J. Phys. F: Metal Phys., vol. 12, pp. 1123–1137, 1982.
29. Gunnarsson, O., "Band model for magnetism of transition metals in the spin-density-functional formalism," J. Phys. F: Metal Phys., vol. 6, pp. 587–606, 1975.

A STUDY OF $Zn_xCd_{1-x}Cr_2Se_4$ SPIN GLASSES

A. V. Myagkov and A. A. Minakov

Abstract The relaxation of the remanent magnetization in semiconductor spin glasses was studied in the system $Zn_xCd_{1-x}Cr_2Se_4$ with $x \sim 0.4$. It was found that the remanent magnetization relaxes more slowly in the highly anisotropic spin glasses in this system with $x \sim 0.4$ created by doping with $\sim 5\,at\%$ Ag, and that the magnetization in the highly anisotropic spin glasses was an order of magnitude larger than that in the weakly anisotropic spin glasses in this system.

INTRODUCTION

Over the past decade spin glasses have become the object of intense theoretical and experimental research efforts. The spin glass state is established in magnetic systems with a chaotic distribution of ferromagnetic and antiferromagnetic interactions. A spin glass refers to a system of spins without overall ferro- or antiferromagnetic ordering in which the spins are configured ("frozen") in directions that change arbitrarily from spin to spin [1]. The typical spin glass is a weak (1–10%) solution of Fe in Au in which the iron ions are chaotically distributed throughout the entire volume of the material, while the sign and magnitude of the exchange interaction depend on the distance between the iron ions. Another situation is possible where all of the magnetic ions of a certain type are distributed in a strictly periodic fashion at the crystalline lattice sites, while the magnitude and sign of the exchange interaction between them are arbitrary. This case is described by the "arbitrary bond" model, which is currently the most advanced model [2]. However, the vast majority of the known spin glasses (AuFe, $Eu_xSr_{1-x}S$, etc.) cannot formally be treated using this model because of the chaotic spatial distribution of the magnetic ions. A unique position in this sense is occupied by the $Zn_xCd_{1-x}Cr_2Se_4$ spin glasses.

$Zn_x Cd_{1-x} Cr_2 Se_4$ monocrystals are ferromagnetic for $x < 0.3$, antiferromagnetic for $x > 0.5$, and spin glasses for $x \sim 0.4$ [3, 4]. The magnetic subsystem of these spin glasses is made up of identical Cr^{3+} magnetic ions distributed periodically at the octahedral sites in the spinel crystalline structure, and the random distribution of the diamagnetic Cd^{2+} and Zn^{2+} ions produces spatial fluctuations in the signs and magnitudes of the interactions between the Cr^{3+} $3d$ ions. This magnetic system is precisely that envisioned in the "arbitrary bond" model.

Moreover, $Zn_x Cd_{1-x} Cr_2 Se_4$ spin glasses have a cubic magnetocrystalline anisotropy which may be smoothly varied from 10^3 to 10^5 erg/cm^3 by doping with small quantities of silver (Ag $\sim$ 5mol%) [5]. Thus, by studying samples with various Ag concentrations, we can experimentally determine the effect of magnetocrystalline anisotropy on the properties of the spin glasses. This type of work is of particular interest, since recent theoretical studies [6, 7] indicate that the Heisenberg magnetic model cannot describe the properties of spin glasses without taking some form of anisotropic interaction into account. It turns out that the spin glass state can only become stable only when the Hamiltonian for the system contains a term that takes some type of anisotropic interaction into account (for example, Soukolis, Grest, and Levin [7] discussed a Hamiltonian with the anisotropic Dzyaloshinsky-Moriya effect $\sum D_{ij}[\mathbf{S} \times \mathbf{s}_j]$ as well as the uniaxial anisotropy $\sum D(S_i^z)^2$). Only under this condition can the characteristic properties of spin glasses (such as the maximum in the temperature dependence of the susceptibility [6], as well as the magnetic viscosity and irreversibility effects [7]) be described.

At the same time, any spin configuration is degenerate with respect to rotation angle about the crystallographic axes (i.e., the energy of the system remains unchanged if each spin is rotated by the same angle about an arbitrarily-selected axis) when the anisotropic interaction is zero, since the Heisenberg interaction $J_{ij} \mathbf{S}_i \mathbf{S}_j$ is isotropic. In the presence of anisotropic interactions, this degeneracy is lifted and certain spin configurations become more advantageous from the viewpoint of energy, i.e., metastable states arise. The energy barrier separating these states will clearly depend on the energy of the anisotropy. Consequently, quantitative characteristics of spin glasses (such as the "freeze-out" temperature T_f, etc.) also depend on the anisotropy energy.

Myagkov, Minakov, and Rudov [8] have studied the dynamic magnetic susceptibility of two types of spin glasses:

1) $\sim$ 5% Ag-doped $Zn_x Cd_{1-x} Cr_2 Se_4$ samples with $x = 0.46$ having a cubic magnetocrystalline anisotropy with a first anisotropy constant $K1 \sim 10^5$ erg/cm^3;

2) undoped $Zn_x Cd_{1-x} Cr_2 Se_4$ samples with $x = 0.4$ and $K1 \sim 10^3$ erg/cm^3.

The paramagnetic Curie points turned out to be similar in the Type 1 and Type 2 samples: 85 and 79 K, respectively. The "freeze-out" temperatures vary by nearly factor of 2, however. For the strongly anisotropic Type 1 samples,

$T_f \sim 37\,\mathrm{K}$, while $T_f \sim 21\,\mathrm{K}$ for the weakly anisotropic Type 2 samples. This difference in "freeze-out" temperature probably indicates that the magnetocrystalline anisotropy itself affects T_f and that T_f indeed increases as the magnetocrystalline anisotropy increases. This result is in qualitative agreement with the theoretical results in [6, 7].

One of the characteristic properties of spin glasses is magnetic viscosity. In the present study, we also studied the relaxation of the remanent magnetization in the same strongly anisotropic (Type 1) and weakly anisotropic (Type 2) $Zn_x Cd_{1-x} Cr_2 Se_4$ samples studied in [8].

1. METHOD USED TO MEASURE THE REMANENT MAGNETIZATION

The remanent magnetization was measured using a rotating-sample magnetometer whose principle of operation involves measuring the voltage induced in the detection coils of the magnetometer as the sample is rotated. If the sample has a remanent magnetic moment, a voltage proportional to the remanent magnetization will be induced in the detector coils at the sample rotation frequency ω_{rot}. One feature of this technique for measuring the remanent magnetization is that the remanent magnetization of the sample can be measured even in the presence of an external magnetic field. Indeed, the magnetic moment of the sample $\mathbf{M}$ in a field $\mathbf{H} \neq 0$ may be written in the following form:

$$\mathbf{M} = \mathbf{M}_r + \mathbf{M}(\mathbf{H}),$$

where $\mathbf{M}_r$ is the remanent magnetic moment of the sample, and $\mathbf{M}(\mathbf{H})$ is the magnetic moment induced by the external magnetic field $\mathbf{H}$. As the sample is rotated, the remanent magnetic moment rotates with it, while the direction of the magnetic moment $\mathbf{M}(\mathbf{H})$ is either identical to that of the external field or oscillates with frequency $2n\omega_{\mathrm{rot}}$ $(n = 1, 2, \ldots)$ if the sample is anisotropic and is being rotated around an axis of the nth order. Therefore, by measuring the voltage at frequency ω_{rot}, we can determine the remanent magnetization of the sample itself, even in the presence of an external magnetic field, i.e., study the effect of a rotating field on the relaxation of the remanent magnetization. Note that traditional methods (for example, a vibration magnetometer) only allow measurement of the remanent magnetic moment of a sample in zero external magnetic field.

The sample rotation frequency was 18 Hz, and the signal was measured using an amplifier with a UPI-1 synchronous detector. An obturator placed between a photodiode and a light-emitting diode and fastened to the rod used to rotate the sample was used to generate a reference signal. The detector coils and the sample were housed in a helium cryostat. A temperature stabilization system made it possible to carry out measurements over the temperature range from

4.2 to 300 K. A laboratory electromagnet with a maximum field strength of up to 10 kOe was used to create the external magnetic field. The sensitivity of the magnetometer to the sample magnetic moment was $\sim 10^{-4}$ G $\cdot$ cm^3.

Rotating-sample magnetometers can be used to determine the corresponding magnetocrystalline anisotropy constants by studying the voltages at frequencies that are even multiples of the sample rotation frequency. In the present work, we are studying $Zn_xCd_{1-x}Cr_2Se_4$ samples (for which Minakov and Filatov have determined the appropriate cubic magnetocrystalline anisotropy constants [5]).

2. THE REMANENT MAGNETIZATION OF $Zn_xCd_{1-x}Cr_2Se_4$ SPIN GLASSES

In this section, we present the results of a study of the remanent magnetization of highly anisotropic (Type 1) and weakly anisotropic (Type 2) $Zn_xCd_{1-x}Cr_2Se_4$ spin glasses using a rotating-sample magnetometer. Since the remanent magnetization of the spin glasses and the rate of relaxation of the remanent magnetization depend on different parameters of the magnetization process, we must identify how the remanent magnetization was created. Two types of remanent magnetization were considered:

TRM (4.2 K, H) - the remanent magnetization of the sample after cooling in a static magnetic field H from a temperature greater than the "freeze-out" temperature to 4.2 K.

IRM (T, H, t_H) - the remanent magnetization of the sample after isothermal magnetization by the field H for time t_H at temperature T. After isothermal magnetization, the sample was either cooled from a temperature greater than the "freeze-out" temperature to the temperature T in zero field or demagnetized at temperature T in a rotating magnetic field with the field strength decreasing continuously from 7 kOe to zero over a time of approximately 1 min.

2.1. DEPENDENCE OF THE REMANENT MAGNETIZATION ON THE FIELD USED FOR MAGNETIZATION

Fig. 1 shows the TRM (4.2 K) and IRM (4.2 K) as a function of field strength H for samples of both types. The TRM (4.2 K) as a function of H shows a maximum, while the IRM (4.2 K) is a saturation curve as a function of H. The functions shown in Fig. 1 are characteristic of spin glasses. The saturated remanent magnetization M_{rs} at $T = 4.2$ K and $t = 50$ s is 40 G in the Type 1 samples and approximately 3.5 G in the Type 2 samples, where t is the time that has passed since the external magnetic field was turned off. Note that the magnetization at saturation M_s is approximately 380 G for both the weakly anisotropic and highly anisotropic samples.

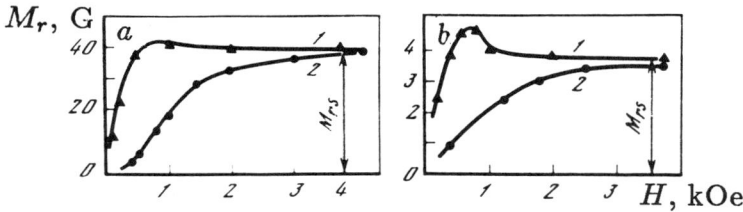

Figure 1   TRM (4.2 K) and IRM (4.2 K, $t_H = 30\,$s) plotted as a function of field strength at $t = 50\,$s for the Type 1 samples (a) and the Type 2 samples (b). 1) TRM, 2) IRM.

2.2. REMANENT MAGNETIZATION RELAXATION LAW FOR $Zn_x Cd_{1-x} Cr_2 Se_4$ SPIN GLASSES

Research on the relaxation of remanent magnetization at 4.2 K indicates [9] that the relaxation of the remanent magnetization in the two types of samples (in which the magnetization was generated by two different techniques (TRM or IRM)) in various magnetizing fields **H** and with different lengths of time for magnetization (for IRM) is described by a power law, where the exponent α depends on the parameters of the magnetization process. Thus,

$$M_r = M \exp\left(-\alpha \ln \frac{t}{\tau}\right), \qquad (1)$$

where M and τ are parameters having dimensions of magnetization and time, respectively.

It turned out that at sufficiently high fields $H \alpha_{\mathrm{IRM}} \approx \alpha_{\mathrm{TRM}}$, α_{TRM} decreases monotonically as the field decreases, while α_{IRM} increases (Fig. 2). Note that under identical conditions, the exponent α is significantly smaller in the Type 1 samples, i.e., the remanent magnetization relaxes more slowly in the highly anisotropic samples. The values exponent α at which the remanent magnetization saturates are ~ 0.05 and ~ 0.15 for the Type 1 and Type 2 samples, respectively.

For magnetization fields smaller than the field at which the remanent magnetization saturates, the IRM and the corresponding exponent α_{IRM} depend on the duration of the magnetization: the IRM increases with increasing t_H, while α_{IRM} decreases with increasing t_H (Fig. 3).

It should be noted that it is impossible to simultaneously determine all of the parameters in the relaxation law $(M, \tau, \text{and } \alpha)$ from the relaxation of the remanent magnetization. Equation (1) implies that $\ln M_r = \ln M\tau^\alpha - \alpha \ln \tau$, i.e., that we can determine the exponent α and the quantity $\ln M\tau^\alpha$ by plotting the experimental relation $M_r(t)$ on a log–log plot; however, the quantity $\ln M\tau^\alpha$ cannot be used to determine the parameters M and τ separately. Some additional information is therefore needed to determine M and τ.

In the present work, we used the following relation obtained from (1) to determine the parameters of the relaxation law:

$$\ln M_r(t) = \ln M - \alpha \ln(t/\tau) \tag{2}$$

We shall assume that only the exponent α changes when the magnetization conditions change. If we determine the quantity $M_r(t_0)$ at fixed t_0 as well as the corresponding values of α and we then plot a graph of $\ln M_r(t_0)$ versus α, it equation (2) implies that the experimental points will lie on a straight line whose slope can be used to determine $\ln(t_0/\tau)$ and, thus, the value of τ, while the y intercept yields the value of $\ln M$. Thus, in order to determine the parameters M and τ, it was necessary to establish under what conditions $\ln M_r(t_0)$ as a function of α at fixed t_0 is linear, and then use this linear section to obtain the values of the parameters M and τ under these conditions of magnetization.

2.3. EXPERIMENTAL DETERMINATION OF THE PARAMETERS M AND τ FOR A TYPE 1 SAMPLE

The function $\ln M_{r_s}$ ($t = 50\,\text{s}$) as a function of α_{r_s} turned out to have a well-defined linear section running from 4.2 K to ~ 10 K (Fig. 4). This makes it possible to determine the parameters $M_0 \approx 120\,\text{G}$ and $\tau_0 \sim 10^{-7}\,\text{s}$ (the subscript "0" indicates that these parameters are associated with the saturated remanent magnetization). On this temperature range, the exponent α_{r_s} is directly proportional to the temperature , i.e, $\alpha_{r_s} \sim T/T_0$, where $T_0 \sim 80$ K.

Thus, on the temperature range from 4.2 to 10 K, the relaxation law for the saturated remanent magnetization of a type 1 sample takes the following form in

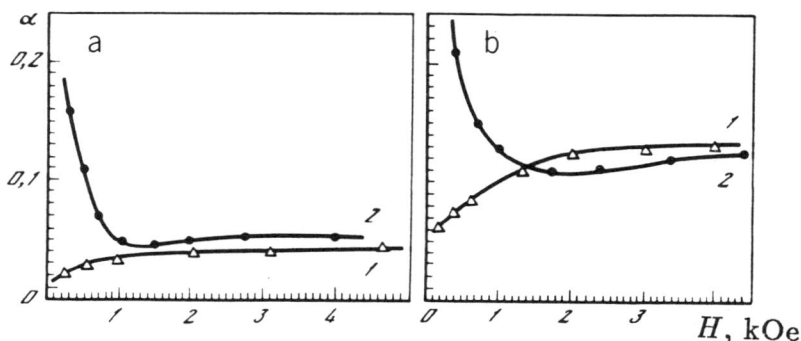

Figure 2   The exponent α corresponding to the relaxation of M_r at 4.2 K as a function of the value of the magnetization field H for TRM (1) and IRM (2) with $t_H = 30\,\text{s}$. a) Type 1 sample; b) Type 2 sample.

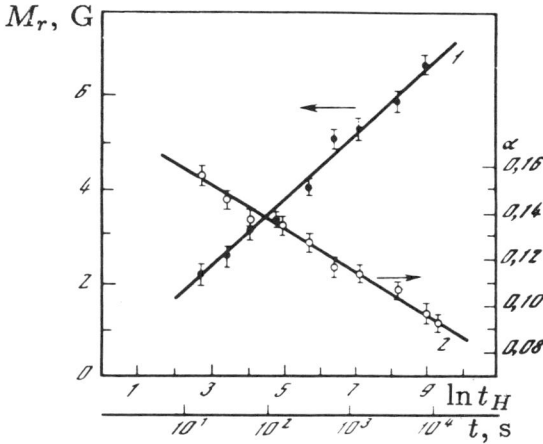

Figure 3 M_r and α as a function of duration of magnetization t_H for a Type 1 sample. 1) IRM (4.2 K, $H = 400$ Oe) at $t = 50$ s; 2) exponent α corresponding to IRM (4.2 K, $H = 400$ Oe).

the CGS system:

$$M_{rs} = 120 \exp\left(-\frac{T}{80} \ln \frac{t}{10^{-7}}\right).$$

It also turned out to be possible to determine the parameters of the relaxation law for the saturated isothermal remanent magnetization. Fig. 5 shows the experimental graph of $\ln M_r$ ($t = 50$ s) versus α for IRM (4.2 K, $t_H = 30$ s) under various magnetization fields and for IRM (4.2 K, $H = 400$ Oe) under various durations of magnetization. This relation has a linear section on which the IRM (4.2 K, $t_H = 30$ s) points lie for $H < 900$ Oe, along with the points corresponding to IRM (4.2 K, $H = 400$ Oe) for any duration of magnetization. The value of τ determined from this linear section for nonsaturated isothermal remanent magnetization is identical to τ_0, and is equal to $\sim 10^{-7}$ s, while the parameter M for the unsaturated isothermal remanent magnetization is approximately equal to 50 G, which is a factor of 2.4 smaller than M_0. Note that it is also possible to determine M_0 using the relation shown in Fig. 5; this value is in agreement with that determined above.

2.4. EXPERIMENTAL DETERMINATION OF THE PARAMETERS M AND τ FOR A TYPE 2 SAMPLE

Fig. 6 shows lnM_r ($t = 50$ s) as a function of α for the nonsaturated isothermal

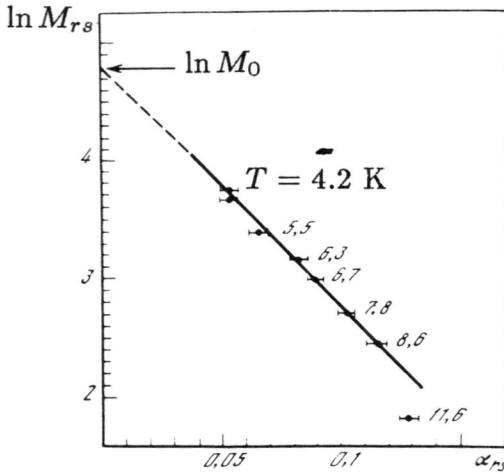

Figure 4 $\ln M_{rs}$ $(t = 50\,\text{s})$ as a function of α_{rs} for a Type 1 sample.

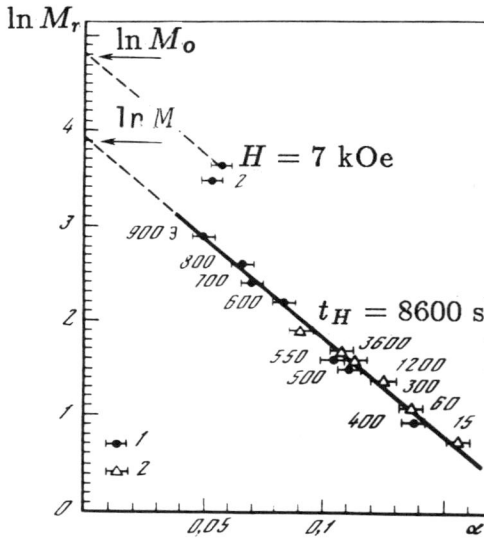

Figure 5 $\ln M_r$ $(t = 50\,\text{s})$ as a function of α for a Type 1 sample. 1) IRM $(4.2\,\text{K}, t_H = 30\,\text{s})$; 2) IRM $(4.2\,\text{K}, H = 400\,\text{Oe})$.

remanent magnetization of a Type 2 sample. The values of the parameters determined using this relation are $\tau \sim 10^{-4}$ s, $M \sim 12$ G, $M_0 \sim 30$ G. Note that the ratio $M_0/M \approx 2.25$ for a Type 2 sample; these values are similar to the values of this ratio for a Type 1 sample.

3. DISCUSSION OF THE RESULTS

Today, most researchers believe that remanent magnetization relaxation in spin glasses is described by a power law [10]. However, the difficulty outlined above in determining the parameters of the power law have resulted in a situation where (despite extensive research on the relaxation of spin glasses) the parameters of the relaxation law have only been determined for saturated remanent magnetization (see, for example, Berton, Chassy, and Odin [11]).

The present study was the first in which all of the parameters of the relaxation law for the unsaturated remanent magnetization of a spin glass were successfully determined. In our view, it is quite surprising that the parameter M is independent of the magnitude of the magnetizing field for unsaturated IRM. This is probably a result of the fact that a spin glass is a cooperative state involving all spins in the system. It can be shown that the parameter M is proportional to the magnitude of the magnetization field in the hard noninteracting cluster model.

The results obtained here also indicate that highly anisotropic $Zn_x Cd_{1-x} Cr_2 Se_4$ spin glasses have a higher "magnetic viscosity" in the sense that their remanent magnetization is much larger, and relaxes much more slowly than the remanent magnetization of weakly anisotropic spin glasses in this system. The parameter M is larger in the strongly anisotropic samples, while the parameter α, which characterizes the rate of relaxation of the remanent magnetization is smaller for the highly anisotropic samples.

Taking the fact that the test spin glasses have an identical crystalline structure, are similar in chemical composition and have similar paramagnetic Curie points [8] into account, the cubic magnetocrystalline anisotropy can be regarded as being specifically responsible for the fact that the "freeze-out" temperature is higher, the remanent magnetization is greater, and the remanent magnetization relaxes more slowly in the highly anisotropic $Zn_x Cd_{1-x} Cr_2 Se_4$ spin glasses [8] than in weakly anisotropic spin glasses of the same system.

This conclusion is in qualitative agreement with the results of Walstedt and Walker [6] and Soukolis, Grest, and Levin [7], where numerical methods were used to study the properties of spin glasses as a function of the anisotropic interaction energy within the framework of the Heisenberg "arbitrary bond" model.

Thus, the results of the present study indicate the that it is important to take the anisotropic interactions into account in describing the spin glass state.

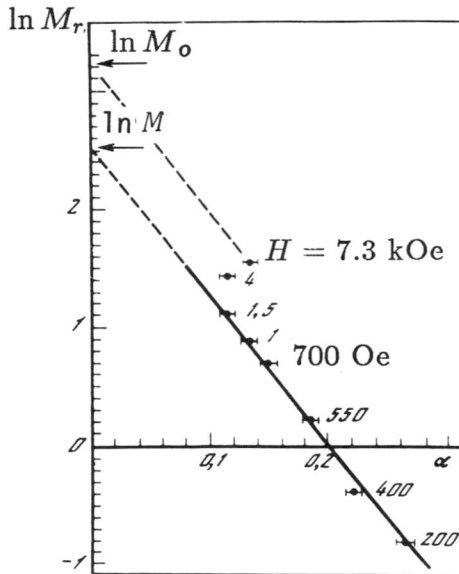

Figure 6   ln M_r (50 s) as a function of α for a Type 2 sample for IRM (4.2 K, $t_H = 30$ s).

REFERENCES

1. Edwards, S. F., and P. W. Anderson, "Theory of spin glasses," J. Phys. F: Metal Phys., vol. 5, pp. 965–974, 1975.
2. Binder, K., and W. Kinzel, "Spin glass model with short-range interactions: A short review of numerical studies," Lect. Notes Phys., no. 149, pp. 124–144, 1981.
3. Minakov, A. A., and V. E. Makhotkin, "The effect of doping on the magnetic structure of semiconductor spinel," Izv. Akad. Nauk SSSR, Ser. Fiz., vol. 44, no. 7, pp. 1473–1479, 1980.
4. Sadykov, R. A., P. L. Grugdin, A. A. Minakov, et al., "Neutron diffraction investigations of $Zn_x Cd_{1-x} Cr_2 Se_4$ magnetic superconductors," Pis'ma v Zh. Éksp. Teor. Fiz., vol. 28, no. 9, pp. 596–599, 1978.
5. Minakov, A. A., and A. V. Filatov, "The magnetic structure of a $Zn_x Cd_{1-x} Cr_2 Se_4$ semiconductor and the effect of doping on this structure at critical concentrations x," In: II seminar po amorfnomu magnetizmu (25–27 iyunya 1980), tez. dokl. [The second seminar on amorphous magnetism (25–27 June, 1980), Abstracts] Krasnoyarsk, 1980, p. 129.

6. Walstedt, R. W., and L. R. Walker, "Monte Carlo simulation of a spin-glass transition," Phys. Rev. Lett., vol. 47, no. 22, pp. 1624–1627, 1981.

7. Soukoulis, C. M., G. S. Grest, and K. Levin, "Absence of irreversibility in isotropic Heisenberg spin-glasses: Anisotropy effects," Phys. Rev. Lett., vol. 50, no. 1, pp. 80–83, 1983.

8. Myagkov, A. V., A. A. Minakov, and S. G. Rudov, "Investigation of the susceptibility of $Zn_xCd_{1-x}Cr_2Se_4$ spin glasses," Preprinty Fiz. Inst. Akad. Nauk SSSR (Lebedev Phys. Inst.), no. 224, 1983.

9. Veselago, V. G., A. A. Minakov, and A. V. Myagkov, "Investigation of the relaxation processes of the remanent magnetization of $Zn_xCd_{1-x}Cr_2Se_4$ spin glasses with cubic magnetocrystalline anisotropy," Pis'ma v Zh. Éksp. Teor. Fiz., vol. 38, no. 5, pp. 255–257, 1983.

10. Maletta, H., "Distinction of spin-glass freezing from superparamagnetic blocking," J. Magn. and Magn. Mater., vol. 24, pp. 179–185, 1981.

11. Berton, A., Chassy, J., Odin, J., et al., "Apparent specific heat of spin glass (AuFe 6 at%) in presence of remanent magnetization and associated energy and magnetization relaxation," Solid State Comm., vol. 37, no. 3, pp. 241–245, 1981.

SUBJECT INDEX

213

215